Stink Bugs
OF
Economic
Importance
IN
America
North
OF
Mexico

COVER PHOTOGRAPHS

Front Cover:
(top to bottom)

Photos 1 and 3: Nymphs and adult, respectively, of the green stink bug, *Acrosternum hilare* (Say).
Photo 2: Adult of the brown stink bug, *Euschistus servus* (Say).
Photo 4: Eggs of *Euschistus* sp.
(All photographs by Robert M. McPherson.)

Stink Bugs OF Economic Importance IN America North OF Mexico

J.E. McPherson
Department of Zoology
Southern Illinois University
Carbondale, Illinois

R.M. McPherson
Department of Entomology
University of Georgia
Coastal Plain Experiment Station
Tifton, Georgia

CRC Press
Boca Raton   London   New York   Washington, D.C.

Library of Congress Cataloging-in-Publication Data

McPherson, J. E. (John Edwin), 1941-
    Stink bugs of economic importance in America north of Mexico / by J.E. McPherson
and R.M. McPherson.
        p. cm.
    Includes bibliographical references and index (p.).
    ISBN 0-8493-0071-1 (alk. paper)
        1. Pentatomidae—United States--Identification. 2.
Pentatomidae—Canada--Identification. 3. Agricultural pests—United
States—Identification. 4. Agricultural pests—Canada—Identification. I. McPherson, R. M.
II. Title
SB945.P47 M37 2000
632'.7549—dc21   00-031279

© 2000 by CRC Press LLC

No claim to original U.S. Government works
International Standard Book Number 0-8493-0071-1
Library of Congress Card Number 00-031279
Printed in the United States of America 1 2 3 4 5 6 7 8 9 0
Printed on acid-free paper

Foreword

Phytophagous pentatomids are pests of a multitude of crops in North America and frequently must be dealt with when designing integrated pest management programs (IPM). This book represents an integration of expertise by two entomologists with vast experience with this important insect family. One has devoted much of his career to studies of the systematics and biology of stink bugs, whereas the other has conducted research for nearly two decades on the management of the pest species on several major commodities, especially in the southern United States where environmental conditions are favorable.

The literature on pestiferous pentatomids is as widespread as their host range. This volume brings together that literature wherein the reader can find information on systematics, biology, behavior, ecology, distribution, injury, damage, economic importance, and management strategies by opening one book from the shelf. All of these aspects must be understood to develop IPM systems for sustainable crop production.

The book is timely because, as the authors indicate, the advent and adoption of transgenic crops have highlighted the importance of pentatomids on some major crops. Frequently suppressed by pesticide applications for key and secondary pests, stink bugs have become troublesome on transgenic crops (e.g., Bt-cotton, when pesticide loads are removed or reduced). Whether pentatomids will achieve pest status on these crops remains debatable, but applied entomologists who will make this determination would be well-served to scan the literature cited in the book. Likewise, those involved in IPM on crops such as soybean, where stink bugs are an annual problem, should find the section on *Euschistus* spp. especially interesting. Brown stink bugs commonly are found on various field and orchard crops but often are overshadowed in abundance by the southern green stink bug. Surely, misidentification, underestimation of damage, and sporadic research efforts have been characteristic of the species in this genus. There is a growing awareness that brown stink bugs are emerging as pests, and this book can provide the IPM practitioner who must deal with them a solid base from which to plan management strategies.

It is fortuitous and fortunate that two entomologists who share the same surname, although not immediately related, would find this family of insects worthy of study and ultimately join to combine their knowledge and experiences into this volume. They have compiled an excellent reference.

David J. Boethel
Department of Entomology
Louisiana Agricultural Experiment Station
Louisiana State University Agricultural Center
Baton Rouge, Louisiana

Foreword

Pentatomidae is among the four largest families of Heteroptera, itself (with some 60,000 species) one of the largest groups in Hemimetabola. Most pentatomids feed on plants; some are monophagous (or oligophagous) and many are polyphagous, so within the family there are both specialized and general pest species. Of the estimated 4,570 species worldwide, 222 occur in North America, north of Mexico, where they display a variety of habits, food plants, and size and structural adaptations.

Pentatomids (or stink bugs, so called because of the spicy scent exuded when they are handled) often feed on the reproductive parts of plants: flowers, seeds, fruits. So also do humans, and so pentatomids, like many of their relatives in other families, compete directly with humans for their food. Other stink bugs feed on leaves or stems, and their importance then lies in their weakening of the plants.

Much has been published on the biologies and ecology of North American Pentatomidae; indeed, more has been published on the North American fauna than on that of anywhere else. Yet despite the group's interest, its economic importance, and the wealth of knowledge available about it, we lack any detailed attempt at a synthesis of that knowledge. This book provides such a synthesis for many economically important species of the Nearctic (north of Mexico).

This book synthesizes and analyzes in detail the lives of eight genera and their economically important species. But the book does more than this. Much of the information on these species can be extrapolated to other species in North America and, with care, even to species in other parts of the world. Knowing how most pentatomids discussed here overwinter, one can suggest how pentatomids elsewhere survive unfavorable times; knowing the feeding habits of the pentatomids here, one can surmise the habits of stink bugs elsewhere. The book thus presents hypotheses that can be tested on local stink bugs by entomologists anywhere, and whose testing may prove useful in the control of harmful pentatomids throughout the world. Indeed, one of these North American species, the southern green stink bug, is a major pest nearly everywhere in the world; the information about it in this book, although primarily restricted to North American populations, is certain to be valuable to those combating this pest on other continents.

This book is important for anyone interested in heteropteran biology and ecology, and particularly important to anyone concerned with the damage these stink bugs can do.

Carl W. Schaefer
Department of Ecology and Evolutionary Biology
University of Connecticut
Storrs, Connecticut

Preface

Stink bugs (family: Pentatomidae) occur throughout North America and include both phytophagous and predaceous species. Most are phytophagous, however, and several cause annual economic losses to a wide array of fruits, nuts, row crops, and vegetables. Stink bug-induced injury to the agricultural commodities of America north of Mexico contributes significantly to the economy of the region. For example, losses from stink bugs on soybeans in the southeastern United States (including costs of control and damage) can exceed $60 million in some years. Stink bug feeding results in lower quality of commodities, reduced yields, and, in severe cases, the unacceptability of the damaged crop into the marketplace.

Because so many agricultural products are affected by stink bug feeding, and because stink bug pests also feed on hundreds of weed hosts and uncultivated plants during their life cycles, it is difficult to obtain an overall appreciation of the impact these arthropods have on agriculture throughout the world. *Stink Bugs of Economic Importance in America North of Mexico* presents the first compilation of research reports of economic injury caused by stink bugs on all the major agricultural commodities produced in North America. We have summarized the enormous volume of scientific articles on stink bugs (approximately 700 references cited) into a single, concise book that includes an overview of the major crops attacked, keys to the pest stink bugs, a discussion of the biology of these pests, and a consideration of the management tactics used in the control of these bugs.

Over 200 host plant species have been tabulated, based on reports of stink bug feeding and development noted or implied in the research articles cited. Each of these host plants is alphabetized within three tables by common name, scientific name, and within family, simplifying review of host plant information. In addition, 12 major crops or crop groups (fruit) that are attacked by stink bugs are covered, including historical crop development, production statistics in North America, crop phenology and growth, crop stages attacked, and a description of the crop injury resulting from stink bug feeding.

Correct identification of pest stink bugs is critical to the development of a strategy to manage or control these pests. This book provides comprehensive keys for identification of the 45 species and subspecies of pest stink bugs that commonly are encountered throughout North America, without requiring the use of other published keys to complete the identification process. It includes 64 line drawings that illustrate the morphological characters needed for identification, making it useful even for individuals not trained in stink bug taxonomy.

Stink Bugs of Economic Importance in America North of Mexico presents eight chapters on the most important stink bug species. These chapters highlight life histories (mating, development of immatures, natural enemies, host plants attacked, scent gland secretions, and pheromones), laboratory studies, descriptions of life stages, and a crop-by-crop discussion of economic damage. Twenty-six photographs are included as examples of crop damage and the egg, nymphal, and adult stages.

The final chapter presents an overview of the tactics being used in many different cropping systems to manage stink bug pest infestations. Consequences of using no controls when economic injury levels are reached are discussed. A review of the biological control of stink bug pests is presented. Seven different cultural control practices are highlighted. The benefits and risks associated with chemical control options also are presented. The pest management discussion concludes with a philosophical futuristic outlook on stink bug biology, management, and control.

Stink Bugs of Economic Importance in America North of Mexico provides the first comprehensive treatment of the basic and applied research information on stink bug pests and inclusive taxonomic keys under one cover.

We have compiled a vast amount of information and resources from many agroecosystems, and, therefore, hope the book will be useful to a wide audience. We thank all individuals (listed in the Acknowledgments) who helped us prepare this publication but especially are grateful to Angelika P. Schmid-Riley for her commitment to excellence in preparing the illustrations used in the taxonomic keys and to Duncan K. McClusky, Kathleen G. Fahey, and Kimbra C. Frost for their many hours of library service dedicated to this project. Finally, we express our sincere appreciation for the patience, love, and support provided by our wives Jean M. McPherson and Shelby C. McPherson during the extended period we were preparing this book.

About the Authors

J. E. McPherson is a professor of zoology at Southern Illinois University, Carbondale (SIU). He obtained his Ph.D. in entomology from Michigan State University in 1968 and joined SIU in 1969 as assistant professor. He was promoted to associate professor in 1974 and to professor in 1979.

Dr. McPherson has written broadly on the ecology and systematics of the Heteroptera, particularly the Pentatomoidea, Reduvioidea, and various aquatic and semiaquatic taxa. He is the author or co-author of more than 150 refereed journal articles, has presented papers at both national and regional meetings, and has given lectures at various universities. He has received several research grants, primarily from the USDA Forest Service.

Dr. McPherson is a member of the Entomological Society of America (ESA) and has served on numerous ESA national and branch committees. He was the recipient of an ESA National Service Award in 1991 for his work on the Editorial Board of the *American Entomologist*, the society's magazine; he currently serves as editor of that publication. He was the 1993 recipient of the ESA North Central Branch's Distinguished Achievement Award in Teaching and the Branch's 1997 recipient of the C. V. Riley Achievement Award. He also was the 1996 recipient of the Outstanding Teacher in the College of Science, SIU. He is a member of several additional societies including the Entomological Society of Canada, Entomological Society of Washington, Florida Entomological Society, Kansas Entomological Society, Michigan Entomological Society, New York Entomological Society, and the Sigma Xi honor society.

Robert M. McPherson is a professor of entomology at the University of Georgia Coastal Plain Experiment Station in Tifton. He obtained his Ph.D. in entomology from Louisiana State University in 1978. He joined the Virginia Tech University faculty in 1978 as an assistant professor and was promoted to associate professor in 1986. In 1987, he accepted an associate professor position at the University of Georgia and was promoted to professor in 1996.

Dr. McPherson has published in the areas of economic entomology, insect pest management, alternative control strategies, and biological control of arthropod pests associated with soybeans, tobacco, small grains, and cotton. He is the author or co-author of more than 70 refereed articles, 50 abstracts/proceedings, 20 experiment station publications, 105 Arthropod Management Test Reports, 70 extension publications, and 90 other scientific articles. He has made numerous presentations at state, regional, national, and international scientific meetings and conferences. He has received research grants from USDA-NAPIAP, Philip Morris USA, R.J. Reynolds, Brown & Williamson, Potash & Phosphate Institute, Soybean/Tobacco/Corn Commodity Commissions, and numerous agricultural pesticide companies.

Dr. McPherson is a member of the Entomological Society of America and was the recipient of the Eastern Branch ESA Distinguished Achievement Award in Extension in 1984; he has also served on the ESA Thomas Say Publications Editorial Board, the Insect Survey and Loss Committee, and the Membership Committee as well as on numerous ESA Branch Committees. Dr. McPherson also is a member of the Georgia Entomological Society; the Sigma Xi honor society, serving as president of the Tifton Chapter in 1998–1999; and the American Soybean Association, receiving the ASA Researcher Award in 1985 and a Meritorious Service Award from The Virginia Soybean Association in 1986.

Acknowledgments

Several individuals contributed their time and expertise to the completion of this book.

David J. Boethel (Department of Entomology, Louisiana Agricultural Experiment Station, Louisiana State University Agricultural Center, Baton Rouge) and Carl W. Schaefer (Department of Ecology and Evolutionary Biology, University of Connecticut, Storrs) reviewed the final draft of the manuscript and prepared one of the Forewords.

The following people reviewed portions of the text and offered helpful suggestions: C. Scott Bundy, James D. Dutcher, G. A. Herzog, and James W. Todd (Department of Entomology, Coastal Plain Experiment Station, University of Georgia, Tifton); G. David Buntin (Department of Entomology, Georgia Experiment Station, University of Georgia, Griffin); Jeffrey R. Aldrich (USDA-ARS, Insect Chemical Ecology Laboratory, BARC-West, Beltsville, MD); David J. Boethel (Department of Entomology, Louisiana Agricultural Experiment Station, Louisiana State University Agricultural Center, Baton Rouge); Lester E. Ehler (Department of Entomology, University of California, Davis); David A. Rider (Department of Entomology, North Dakota State University, Fargo); Michael T. Smith (USDA-ARS-BIIR, Newark, DE); George L. Teetes (Department of Entomology, Texas A&M University, College Station); Donald B. Thomas, Jr. (USDA-ARS, Subtropical Agriculture Research Laboratory, Weslaco, TX); M. O. Way (Texas A&M University, Agriculture Research & Extension Center, Beaumont); and Kenneth V. Yeargan (Department of Entomology, University of Kentucky, Lexington).

James R. Fuxa (Department of Entomology, Louisiana Agricultural Experiment Station, Louisiana State University Agricultural Center, Baton Rouge) and John S. Russin (Department of Plant and Soil Science, Southern Illinois University, Carbondale) provided taxonomic help with fungi classification, and Norman E. Woodley (Systematic Entomology Laboratory, PSI-ARS-USDA, % National Museum of Natural History, Washington, D.C.) provided taxonomic help with tachinid fly classification. Carroll W. Johnson (USDA, Coastal Plain Experiment Station, Tifton) and Michael O. Moore (Department of Biology, University of Georgia, Athens) provided identifications of host plants and verifications of plant scientific names. Ronald H. Cherry (Everglades Research and Education Center, University of Florida, Belle Glade), Joseph E. Eger (Dow Agrosciences, Tampa, FL), Walker A. Jones (USDA, ARS, SARL, BCPRU, Weslaco, TX), and Ronald D. Oetting (Department of Entomology, Georgia Experiment Station, Griffin) provided reprints and photocopies of literature. Cecil L. Smith (Museum of Natural History, University of Georgia, Athens) and C. Scott Bundy (Department of Entomology, Coastal Plain Experiment Station, University of Georgia, Tifton) loaned specimens used to prepare several of the illustrations accompanying the taxonomic keys.

All figures were drawn by Angelika P. Schmid-Riley (Coastal Plain Experiment Station, University of Georgia, Tifton).

We are much indebted to Kathleen G. Fahey and Kimbra C. Frost (Morris Library, Southern Illinois University, Carbondale), and Duncan K. McClusky (Library, Coastal Plain Experiment Station, University of Georgia, Tifton) who went far beyond the call of duty in obtaining interlibrary loans, or photocopies, of many of the cited references.

And, finally, we give a special note of appreciation to secretaries Jenny V. Granberry and Carol S. Ireland (Coastal Plain Experiment Station, University of Georgia, Tifton) for their hard work in getting this manuscript into proper form.

Contents

1 General Introduction to Stink Bugs

CONTENTS

1.1 TAXONOMIC POSITION, BACKGROUND, AND ECONOMIC IMPORTANCE

The family Pentatomidae (stink bugs) occurs worldwide and is comprised of a large group of phytophagous and predaceous species. It is included in the superfamily Pentatomoidea, which in America north of Mexico (hereafter referred to as North America) also includes the Scutelleridae (shieldbacked bugs), Cydnidae (burrower bugs), Thyreocoridae (negro bugs), and Acanthosomatidae (acanthosomatids).

Pentatomoids are members of the order Heteroptera or true bugs (see recent discussion of heteropteran classification by Schuh and Slater [1995]). Heteroptera are characterized by (1) a segmented beak that arises from the front of the head and (2) wings that, when present and well-developed, are folded flat on the abdomen with the forewings usually leathery basally and membranous distally and the hind wings membranous. Pentatomoids usually can be recognized by their 5-segmented antennae and a well-developed scutellum, which sometimes covers the entire abdomen.

Most pentatomoids are phytophagous, the primary exceptions being the predaceous asopine stink bugs. Of the four subfamilies of phytophagous stink bugs that occur in North America (i.e., Discocephalinae, Edessinae, Pentatominae, Podopinae), only the Pentatominae contains species of major economic importance. The Pentatominae includes approximately 40 genera and 180 species and subspecies in this broad geographical area (Froeschner 1988), but in recent years only five have received much attention as major pests [i.e., southern green stink bug, *Nezara viridula* (L.); rice stink bug, *Oebalus pugnax pugnax* (Fab.); green stink bug, *Acrosternum hilare* (Say); brown stink bug, *Euschistus servus* (Say); and onespotted stink bug, *Euschistus variolarius* (Palisot de Beauvois)]. Others [e.g., *Euschistus conspersus* Uhler; dusky stink bug, *Euschistus tristigmus* (Say); conchuela, *Chlorochroa ligata* (Say); and redshouldered stink bug, *Thyanta custator accerra* McAtee] occasionally can cause severe damage; and still others [e.g., Say stink bug, *Chlorochroa sayi* (Stål); and harlequin bug, *Murgantia histrionica* (Hahn)] were considered major pests at one time but, today, apparently are of minor importance.

Pentatomine stink bugs, as a group, feed on a wide range of fruit, vegetable, nut, and grain crops, as well as wild hosts; they are associated most often with "heads" or fruits but will feed on stems and leaves. Although economic species have been reported from numerous plants (e.g., McPherson 1982), research generally has concentrated on their association with only a few of these, which will be considered within the expanded discussion of each stink bug. For example, *Nezara viridula* has been collected from many plants, including several economic species, but most research in North America has concerned its damage to soybean.

Loss in dollars from specific stink bugs is impossible to assess accurately in North America because of the way economic data are gathered (i.e., stink bugs often are not treated as individual species but lumped with other stink bugs or other phytophagous heteropterans) and because there has been little, if any, effort to combine loss data from stink bugs for a specific crop from different states or regions of the country. However, data are available for some states for some crops and, although of limited value, give some appreciation of the potential economic loss from these insects.

Another problem in accurately estimating the cost of stink bug damage for some crops results from changes in agricultural practices in some areas of the country, which has led to an upsurge in stink bug problems. For example, stink bug feeding on cotton has increased in recent years (Bacheler and Mott 1995). Much of this increase has occurred in the Southeast where the boll weevil eradication program is under way (Bacheler and Mott 1995, Lambert and Herzog 1993). Through the late 1970s, the boll weevil, *Anthonomus g. grandis* Boheman, and bollworm, *Helicoverpa zea* (Boddie), were controlled by multiple applications of organophosphate insecticides during the early season. These applications tended, coincidentally, to keep stink bug populations at low levels. However, stink bugs no longer are being controlled because of the reduction and/or elimination of these early-season applications. Currently, loss of cotton bolls, sometimes in excess of 30% of the bolls (Bacheler and Mott 1995), as well as a decrease in mature fibers and in seed germination rate (Barbour et al. 1990), reportedly is the result of stink bug feeding (Bacheler and Mott 1995, Barbour et al. 1990). In 1994, over $7.6 million in crop losses and control costs were reported for stink bugs in the Beltwide Cotton Insect Loss Report (Williams 1995).

For the last several years, Georgia, through its Agricultural Experiment Stations, has prepared annual loss estimates to insects, including stink bugs (see Appendix). For example, loss from stink bugs (including *Nezara viridula*, *Acrosternum hilare*, and *Euschistus servus*) for soybean in 1992 was estimated at $5,646,000 (Hudson et al. 1993) and in 1993 at $2,341,000 (McPherson et al. 1995a). Similarly, 1992 losses from stink bugs (species not indicated) was estimated at $10,657,000 for field corn (All and Hudson 1993), $280,000 for small grain (Hudson and Buntin 1993), $689,600 for tomatoes, $149,700 for sweet corn, $382,700 for snap beans, $328,300 for okra, $43,400 for lima beans (Adams and Chalfant 1993), and $33,000 for grain sorghum (Hudson and All 1993). Loss from *E. servus*, *N. viridula*, various species of leaffooted bugs (Coreidae), and the tarnished plant bug, *Lygus lineolaris* (Palisot de Beauvois) (Miridae), as a group, was estimated at $889,000 for peaches (Horton et al. 1993). Finally, loss from stink bugs and plant bugs, as a group, was estimated at \$27,164,000 (including \$7,276,000 for control and \$19,887,000 from damage) for several crops combined, including apple, cotton, field corn, grain sorghum, peach, pecan, small grain, soybean, and vegetables (McPherson and Douce 1993).

Georgia is not alone in suffering severe economic loss from stink bugs. In Texas, for example, annual loss from *Oebalus p. pugnax* for rice is estimated at \$13,000,000; and from panicle-feeding bugs, including *Nezara viridula*, *O. p. pugnax*, *Chlorochroa ligata*, leaffooted bug [*Leptoglossus phyllopus* (L.)] (Coreidae), and false chinch bug (*Nysius raphanus* Howard) (Lygaeidae) the loss for sorghum is estimated at \$1,900,000 (George Teetes, personal communication).

Granted, these loss values for Georgia and Texas are estimates only, but they do give some appreciation of the enormous potential of stink bugs to cause significant problems in crop production.

1.2 DAMAGE TO PLANTS

Adults and nymphs obtain their food by piercing plant tissues with their mandibular and maxillary stylets and extracting plant fluids. Although all stages except the first instar (a nonfeeding stage) feed on plant material (Simmons and Yeargan 1988a, Todd 1989), adults (Duncan and Walker 1968) and/or fifth instars (McPherson et al. 1979a, Yeargan 1977) cause the most damage, at least to soybean. Stink bugs may attack all parts of the plant including stems, petioles, foliage, flowers and fruit/seeds, although damage usually is confined to the fruiting structures (e.g., Anonymous 1998, Barbour et al. 1988, Hallman et al. 1992, McPherson et al. 1994, Negron and Riley 1987, Panizzi 1997, Reich 1991, Riley et al. 1987, Sedlacek and Townsend 1988, Todd 1981, Townsend and Sedlacek 1986). The feeding punctures that result from the insertion of the stylets form minute discolored spots on the plant (Fery and Schalk 1981; Lye et al. 1988a, b; Miner 1966a, b; Negron and Riley 1987; Todd 1981; Todd and Herzog 1980). A stylet sheath often is left by the bug at the feeding site (Apriyanto et al. 1989a; Bowling 1979, 1980; Bundy et al. 2000; Cronholm et al. 1998; Hall and Teetes 1982a, b; Hall et al. 1983; Hollay et al. 1987; Lye and Story 1988; Lye et al. 1988a, b; Marchetti and Petersen 1984; Negron and Riley 1987; Rice et al. 1985; Simmons and Yeargan 1988a; Viator and Smith 1980; Viator et al. 1982, 1983; Yates et al. 1991). Stylet sheath morphology is shaped by the intracellular spaces; however, in at least some seed feeders, the bugs produce a short sheath through the seed coat and then insert their stylets deeper into the seed without extending the sheath (Miles 1968). Air spaces often are produced after the cell contents are withdrawn, leaving a chalky appearance (Miner 1966a). As the damaged seeds dry, dark areas may surround the stink bug punctures, and the inner membrane of the seed coat may become fused abnormally to the cotyledons (Panizzi and Slansky 1985a). This injury can adversely affect seed germination (Jensen and Newsom 1972, McPherson et al. 1979a, Todd 1981, Todd and Turnipseed 1974), although the location of the puncture is more important than the number of punctures on the seed. A single puncture on the axis of the radicle-hypocotyl of a soybean seed can prevent germination (Jensen and Newsom 1972).

In general, stink bug feeding results in the loss of plant fluids (and turgor pressure), the injection of destructive digestive enzymes, the deformation and abortion of seed and fruiting structures, the predisposition to colonization by patho-genic and decay organisms, and delayed plant maturation (Annan and Bergman 1988; Anonymous 1998; Apriyanto et al. 1989a, b; Daugherty 1967; Daugherty and Foster 1966; Jones and Caprio 1994; Lee et al. 1993; McPherson et al. 1994; Men-ezes et al. 1985a; Michailides et al. 1987, 1988; Negron and Riley 1987; Nilakhe et al. 1981a, b, c; O'Keeffe et al. 1991; Ragsdale et al. 1979; Roach 1988; Russin et al. 1987, 1988a, b). Excessive stink bug damage to soybean can result in foliar retention, delayed plant maturation, and abnormal plant growth (Panizzi and Slansky 1985a). Stink bug feeding induces early maturity of tomatoes and, thus, a reduction in fruit size and weight (Lye et al. 1988a); it also can give the fruit a bitter taste and pithy texture (Callahan et al. 1960). Stink bug-induced injury during early formation of seeds and fruits in several crops can result in both of these fruiting structures being shriveled, undersized, and even aborted (Barbour et al. 1990; Grant et al. 1986a; McPherson et al. 1994; Nilakhe 1981a, b, c; Toscano and Stern 1976a). Feeding injury to more mature seeds and fruit causes less severe damage (Jones and Caprio 1994, McPherson et al. 1994). Significant yield and quality crop losses can occur when damage is heavy (All and Hudson 1993; Apriyanto et al. 1989a, b; Barbour et al. 1990; Castro 1985; Hall and Teetes 1982a, b, c; McPherson et al. 1979a, 1992; Miner and Dumas 1980a, b; Nilakhe et al. 1981a, b, c; Thomas et al. 1974; Todd and Turnipseed 1974; Viator and Smith 1980; Viator et al. 1982, 1983; Yeargan 1977), and the crop can even become unmarketable (Adams et al. 1991, Genung et al. 1964). Stink bug losses just to soybean exceed $60 million in some years in the southeastern United States (Table 1.1).

The population dynamics (Drees and Rice 1990; Funderburk et al. 1990, 1991; Jones and Cherry 1986; Jones and Sullivan 1983; McPherson et al. 1993; Schumann and Todd 1982), seasonal occurrence (Fontes et al. 1994), and spatial distribution (Foster et al. 1986, 1989; Lye and Story 1989) of phytophagous stink bugs have been studied in, or in association with, several crops in North America. Sampling tech-niques to determine when damaging populations are present also have been investi-gated for these pests (Deighan et al. 1985, Elliott et al. 1994, Foster et al. 1989, Merchant and Teetes 1992, Todd and Herzog 1980, Zeiss and Klubertanz 1994).

1.3   GENERAL LIFE HISTORY

The habits of North American pentatomine stink bugs are variable, but generaliza-tions are possible. Adults (and sometimes nymphs) overwinter beneath leaf litter and other ground debris and usually remain inactive during this period in most parts of the country. However, in the more southern parts of their ranges in North America, *Murgantia histrionica* (Brett and Sullivan 1974) and *Nezara viridula* (Todd 1989) will feed in winter during mild temperatures.

Adults emerge in spring as temperatures rise and begin feeding and reproducing on various hosts including grasses, herbaceous vegetation, shrubs, and trees. They are attracted most often to plants with growing shoots and developing seeds or fruits. In fact, they will move from host to host as peak reproduction of earlier hosts passes and that of other plants approaches (e.g., Smith 1996a, b; Todd 1989; Todd and Herzog 1980).

TABLE 1.1
Estimates of Losses (Millions of Dollars)[a,b]

Year	AL	AR	FL	GA	LA	MS	NC	SC	TN	TOTAL
1977	$2.15	$0.15	$4.15	$0.20	$2.00	$0.50	0	$9.30	$0.10	$18.55
1978	1.10	0	4.80	10.50	48.00	0.50	0	13.00	0	77.90
1979	1.00	0	5.70	23.60	25.00	—	0	5.40	0.20	60.90
1980	0.80	0	3.20	0	19.00	4.00	$1.00	0.05	1.00	29.05
1981	0.20	0	4.05	2.70	21.00	2.00	0	0.10	0.00	30.05
1982	0.54	0.50	5.10	23.28	25.00	6.00	0.75	0.15	0	61.32
1983	0.77	0.09	4.79	0.43	40.00	10.00	0.60	0.13	1.20	57.01
1984	0.77	0.50	1.12	0.38	26.00	—	0	0.20	0	28.90

[a] Including costs of control and crop damage from stink bugs on soybeans in the southeastern United States 1977–1984.

[b] Southern green, brown, and green stink bugs combined.

Source: Data from annual insect detection, evaluation, and prediction reports, Insect Detection, Evaluation, and Prediction Committee, Southeast. Branch Entomol. Soc. Am., 1977–1984 reports.

Precopulatory and copulatory behaviors have been reported for several species (McPherson 1982), but comparative studies have been rare. Recently, Drickamer and McPherson (1992) compared mating behavior patterns in six species, five of which are members of the same genus (*Euschistus*).

Mating usually begins with the male antennating various parts of the female's body but eventually concentrating on or near the tip of her abdomen. This may be sufficient to stimulate the female to raise the tip of her abdomen for aedeagal insertion. However, antennating may be combined with, or replaced by, butting, and the male may attempt to raise the female's abdomen with his head. Subsequently, the male will turn approximately 180°, elevate his abdomen, and attempt to insert his aedeagus. If successful, copulation may last for several hours in this end-to-end position, and both individuals may feed during this time; during copulation, the female often will drag the male along. If the female is not receptive, she will kick at him with her hind legs or walk away.

Eggs are laid in clusters on various above-ground parts of the plant, particularly on leaves, and adhere to each other and to the plant by a sticky secretion. They usually are abandoned after oviposition, maternal care of the eggs being uncommon (McPherson 1982).

Stink bugs, as is true of most heteropterans, have five nymphal instars (e.g., McPherson 1982, Todd 1989). First instars are gregarious and remain clustered atop or near the egg shells. Although it is reported they do not feed (e.g., Bowling 1980, McPherson 1982), it has been suggested they acquire symbionts by sucking secretions covering the shells of their hatched eggs (Buchner 1965). The nymphs begin to disperse slightly and feed after the first molt, but aggregation may continue through the third molt (Hokyo and Kiritani 1962, Kiritani 1964a, Panizzi et al. 1980, Todd 1989). This nymphal aggregation up to the fourth instar may provide a measure of protection from predators (see study of first instar *Nezara viridula* by Lockwood and Story [1986a]).

As the bugs pass through the remaining instars, subtle changes occur in the markings, patterns, and numbers of punctures and setae, and development of the wing pads. Wing pads are distinctive enough in the fourth and fifth instars that they can be used to distinguish these instars from each other and from younger instars. The most comprehensive morphological studies of the eggs and nymphs of North American stink bugs are those of Esselbaugh (1946) and Javahery (1994) for eggs, and DeCoursey and Esselbaugh (1962) and Yonke (1991) for nymphs. Bundy and McPherson (2000a) have provided scanning electron micrographs of the eggs of several of these stink bugs.

The number of generations per year in North American stink bugs ranges from one in the North to five in the extreme South (e.g., south Florida).

Stink bugs are attacked by numerous invertebrates, several of which are hymenopteran egg parasitoids (e.g., Girault 1907; Krombein et al. 1979a, b, c; Zeiss and Klubertanz 1994) and tachinid fly parasitoids (e.g., Arnaud 1978, Stone et al. 1965, Zeiss and Klubertanz 1994). Among vertebrates, several species of birds have been reported as predators of stink bugs (e.g., Knowlton 1944).

2 Major Crops Attacked

CONTENTS

2.1 SOYBEAN

The soybean dates back nearly 5,000 years when it [*Glycine soja* (L.)] was discovered growing wild in China (Smith 1994). This wild type was domesticated into cultivated soybean [*G. max* (L.)], which does not grow in the wild. The cultivated soybean reached the United States in the mid-1770s as ballast for returning clipper ships. Soybean grew in importance to U.S. farmers in the 1940s as a field crop for feed (forage) and industrial uses. It emerged as a seed and oil crop in the 1950s and, since the 1970s, has become a major agricultural crop for both domestic and export uses (Smith 1994).

Although soybean production in the United States has increased over much of the 20th century (Kogan and Turnipseed 1987, Turnipseed and Kogan 1976), it has increased most rapidly since approximately 1960, particularly in the southern states (Newsom et al. 1980, Turnipseed and Kogan 1976). For example, in 12 midwestern states where soybean has been produced historically, soybean acreage increased from an average of 14,434,000 acres (5,841,000 ha) between 1950 and 1959 to 30,524,000 acres (12,353,000 ha) in 1976 (Newsom et al. 1980). During the same period, even more dramatically, production increased in 11 southern states from an average of 3,245,000 acres (1,313,000 ha) during 1950 to 1959 to 17,670,000 acres (7,151,000 ha) in 1976. Continued expansion of soybean acreage must occur primarily in the South because of the availability of agricultural land in this area (Newsom 1980). The reasons for the increased demand for soybean are discussed briefly by Smith (1994).

The United States produces more soybean than any other country; in 1996/1997, it produced 49.4% of the world's total supply (USDA Nat. Agric. Stat. 1998). Soybean is grown in 29 adjoining states, from Texas northward to North Dakota, eastward along the Canadian border to Pennsylvania and New Jersey, southward along the Atlantic Coastline to Florida, and westward back to Texas (Am. Soybean Assoc. 1997). In 1997, soybean was harvested from 69.9 million acres (28.3 million ha) and was valued at $17.7 billion (Table 2.1) (USDA Nat. Agric. Stat. 1998). Almost 47% was harvested in four north central states: Iowa (10.4 million acres), Illinois (10.0 million acres), Minnesota (6.7 million acres), and Indiana (5.4 million acres). Seventeen of the states planted more than 1 million acres each, and these states represent all areas of the soybean-producing region (USDA Nat. Agric. Stat. 1998).

Knowledge of soybean physiology is important for understanding how plant stressors, such as insect pests, can disrupt plant growth (Teare and Hodges 1994). Three practical aspects of soybean physiology are important for managing insect pests: (1) soybean cultivar type, (2) the growth stage of soybean development, and (3) maturity group. Two types of soybean are grown in the United States. The indeterminate type cultivars, which are produced mainly in the northern states, have a terminal bud that continues to produce new vegetation during most of the season. These cultivars have influorescences on axillary racemes, giving the plant a sparse but rather even distribution of pods on all branches. The determinate type cultivars, which are produced mainly in the southern states, cease vegetative growth of the terminal bud when it begins to flower. These cultivars have both axillary and terminal racemes and are recognized by a dense cluster of pods at the terminal (Teare and Hodges 1994).

Stages of soybean growth are classified as vegetative or reproductive. Vegetative stages are described from the time the plant emerges from the soil (Fehr and Caviness 1977, Fehr et al. 1971). After the cotyledon stage (noted as VC in the classification index), nodes on the main stem are counted to designate the vegetative, or V, growth stage. The V1 stage plant contains only the unifoliolate nodes with fully developed leaves; V2 contains a fully developed trifoliolate leaf at the node above the unifoliolate nodes; and V3, V4, V5, etc., contain three nodes or more on the main stem with fully developed leaves beginning with the unifoliolate nodes. Reproductive soybean growth stages, or R stages, are based on flowering, pod development, seed development, and plant maturation (Fehr and Caviness 1977, Fehr et al. 1971). The R1 (beginning bloom) and R2 (full bloom) stages may occur simultaneously in determinate varieties because flowering begins at the top of the main stem. The two stages are approximately 3 days apart for indeterminate varieties, with flowering beginning in the lower portion of the main stem and progressing upward (Fehr and Caviness 1977). The R3 stage is the beginning pod stage (pods 3/16 inch long at one of the four uppermost nodes), and the R4 stage is the full pod stage with no seeds present. The R5 stage is the beginning seed stage where the soybean pods are filling with seeds, and the R6 stage is the full seed stage where the pods are filled with full-sized seeds. The R7 stage is the beginning of maturity and the R8 stage is full maturity (Fehr and Caviness 1977, Fehr et al. 1971).

The primary control of soybean flowering and maturity during the growing season is through daylength (Teare and Hodges 1994). Daylength is controlled by

TABLE 2.1
1997 Crop Production and Value Summary of Important Agricultural Commodities Reportedly Damaged by Stink Bugs

Commodity	Total Harvested Acres × 1000	Total Production × 1000	Value (Dollars) × 1000
Alfalfa	23,673	79,242 ton	13,416,721[a]
Apple	—	10,226,600 lb	1,687,974
Corn, field			
Grain	73,720	9,365,574 bu	24,394,072
Silage	5,758	91,903 ton	—
Corn, sweet			
Fresh	222.8	2,258,700 lb	398,279
Proces.	464.2	6,647,080 lb	247,839
Cotton	13,283.5	18,976.9 bales	6,142,346
Dry bean[b]	1,720.2	2,915,600 lb	586,452
Macadamia	19.2	58,000 lb	42,920
Pea (green)	268.9	951,880 lb	136,996
Peach	—	2,651,100 lb	451,202
Pear	—	2,088,000 lb	299,621
Pecan	—	272,100 lb	257,043
Potato	1,325.5	45,991,200 lb	2,604,189
Potato, sweet	83.5	1,302,500 lb	213,171
Rice	3,034	17,889,600 lb	1,728,687
Snap bean	195.5	733,000 ton	129,753
Sorghum			
Grain	9,391	653,106 bu	1,455,717
Silage	310	3,885 ton	—
Soybean	69,884	2,727,254 bu	17,704,722
Sugar Beet	1,427.8	29,874 ton	1,211,001[c]
Tobacco	797.3	1,678,821 lb	3,039,217
Tomato			
Fresh	125.4	3,780,900 lb	1,246,843
Proces.	283.4	19,945,300 lb	605,350
Wheat	63,577	2,526,552 bu	8,611,684

[a] Includes value of all hay crops, including alfalfa.
[b] Includes lima, blackeyed, kidney, pinto, and others.
[c] 1996 value is the most recent available for sugar beet.

Source: USDA National Agricultural Statistics 1998.

latitude, with the longer daylengths in the more northern latitudes resulting in longer periods between emergence and flowering of soybean. Soybean plant breeding has resulted in cultivars that require different daylengths to stimulate flowering. These cultivars fall into 12 maturity groups from 00 to X. The 00 cultivars bloom and mature during the long days in southern Canada and the northern United States. The group X cultivars mature in the tropical latitudes of Central America (Teare and Hodges

1994). Soybean producers tend to plant cultivars of several different maturity groups and also tend to plant at different times during the season. These two practices enable growers to expand the flowering and maturing periods of the crop to facilitate better harvesting.

Soybean is attacked by several species of stink bugs, resulting in extensive damage. The bugs are primarily attracted to soybean that is in the reproductive stages, preferring to feed on the developing seeds. The damage to this crop caused by stink bugs is detailed later in the text.

2.2   COWPEA

Cowpea is an important legume crop throughout the world. Annually, about 1.5 to 2.25 million t (1.5 to 2.25 billion kg) of dry seed are harvested from approximately 12.3 to 18.5 million acres (5 to 7.5 million ha) worldwide (Vigna Crop Germplasm Committee 1996). In the United States, cowpea is grown on about 197,600 acres (80,000 ha) each year as a processing pea (southern pea), as a dry bean (blackeye pea, blackeye bean, field pea), and as a garden crop (field pea, crowder pea, purple hull). Cowpea is grown throughout the southern states from Texas to the Carolinas and Virginia. Georgia, Arkansas, and Tennessee produce the most processing peas, and California and Texas produce the most dry beans (Vigna Crop Germplasm Committee 1996). The processing industry also freezes cowpeas, with 13,690,000 kg (30,118,000 lb) frozen in 1988 (Fery 1990).

The vegetative and reproductive phases of cowpea are similar to those detailed for soybean. Cowpeas vary a great deal at the start and end of the reproductive growth period. Some cultivars begin to flower within 30 days after planting and are ready for harvest 25 days later. Other cultivars require more than 100 days to flower and between 210 and 240 days for the crop to mature (Summerfield et al. 1985). The complete phenology of cowpea is reviewed by Summerfield et al. (1985). The reproductive growth is comprised of overlapping periods of development of individual peas, each lasting approximately 19 days. The longer the reproductive period, the greater the number of peas that mature and the higher the yield (Summerfield et al. 1985).

Stink bugs are not an economic threat to the crop until the pods begin to fill with seeds. Stink bug feeding can result in distorted pods and blemished peas, pod abscission, seed abortion, and reduced yield (Chalfant 1985). Specific information on stink bug damage to cowpea (southern pea) is reported by Fery and Schalk (1981).

2.3   CORN

Corn is truly an American crop. It can be produced in every state of the United States except Alaska (Aldrich et al. 1978). It is a fast-growing crop, maturing in approximately 80–120 days after planting, with hybrid selection, region of the country, and growing conditions influencing the length of the growing season (Aldrich et al. 1978).

In 1997, corn was harvested from more acres (73.7 million acres [29.8 million ha]) than any other crop produced in the United States and was valued at $24.4 billion (Table 2.1) (USDA Nat. Agric. Stat. 1998), also the highest valued crop in the country. Iowa (12.0 million acres), Illinois (11.1 million acres), and Nebraska (8.7 million acres) had the highest acreage of corn harvested in the United States in 1997, but 16 states planted 1 million acres or more (USDA Nat. Agric. Stat. 1998). The United States produces more corn than any other country in the world (USDA Nat. Agric. Stat. 1998).

Stink bugs can infest and damage both the vegetative and reproductive stages of corn. The vegetative stages include VE, emergence of the corn plant before the collar of the first leaf is exposed; V1, collar (yellow flared band appearing at the point of leaf attachment to the sheath) of first leaf exposed; V2, collar of second leaf exposed; V3, V4, etc., up to VT, which is the tasseling stage. The reproductive stages include R1, silking begins from the tip of the husk; R2, blister stage where the kernel resembles a small bubble or blister and contains a clear fluid inside; R3, milk stage where the kernel begins to turn yellow and contains a milky white fluid; R4, dough stage where the kernel now contains a thick fluid that resembles biscuit dough; R5, dent stage where all or most of the kernels have dented and begun to harden off; and finally R6, black layer stage where the kernel forms a black layer where it attaches to the cob and is physiologically mature (Wyffels Hybrids 1998).

2.4 SORGHUM

Wild sorghum dates back prior to 3000 BC (Doggett 1988). Grain sorghum first was carried into North America from West Africa with the slave trade, along with other grain crops. In about 1857, 16 cultivars of sorgo were brought to the United States from Natal. Early sorghum development in North America involved natural hybridization. Deliberate hybridization followed soon thereafter, with some of the earliest crosses being made in 1914. This was followed with the development of extensive hybridization programs (Doggett 1988).

Sorghum ranks fifth in acreage and production among major cereal crops on a worldwide basis behind wheat, rice, corn, and barley (Young and Teetes 1977). World production in the mid-1970s was approximately 52 million t (52 billion kg) produced on some 104 million acres (42 million ha). Slightly more than 50% of sorghum grain is produced in North America (Young and Teetes 1977).

In 1997, grain sorghum was harvested from 9.4 million acres (3.8 million ha) in 18 states, producing 653.1 million bushels, and was valued at approximately $1.5 billion (Table 2.1) (USDA Nat. Agric. Stat. 1998). Just over 6.6 million acres were harvested in Kansas and Texas, collectively; most of the other sorghum-producing states are in the south, southwest, and midwest (USDA Nat. Agric. Stat. 1998). In Texas alone, 3–4 million acres are used annually for sorghum production (Fuchs et al. 1988).

Sorghum grain is used for human consumption and as feed for animals (Doggett 1988). Sorghum has been a vital food source for billions of people, especially in the semi-arid tropics of Africa and Asia. Sorghum is prepared in many traditional ways for human consumption. The grain can be used in the making of leavened and unleavened bread and in the production of beer. It also can be boiled into a porridge. Sorghum stems and foliage are used for pasture, greenchop or hay, and silage; plant bases are used as fuel for cooking, and stems are used to make baskets and fish traps (Doggett 1988).

Sorghum also is used as a feed grain for livestock and poultry. It first is coarsely ground, broken, or soaked by various procedures so that less passes undigested through the animal (Doggett 1988).

In 1997, 17 of the 18 states that produced sorghum for grain also produced sorghum for silage. Silage was harvested from 310,000 acres (125,455 ha), producing 3.9 million tons (3.5 billion kg) (USDA Nat. Agric. Stat. 1998). Kansas, Nebraska, South Dakota, and Texas harvested 74.2% of the national total (USDA Nat. Agric. Stat. 1998). Sorghum often is rotated extensively with cotton and soybean.

Stink bugs feed primarily on sorghum kernels and to a lesser extent on stems, branches, and glumes (e.g., Hall et al. 1983). Both nymphs and adults can damage sorghum, and damaged kernels rarely develop fully and may be lost during harvest (Cronholm et al. 1998). Also, sorghum kernels attacked by stink bugs can become infected with fungi (Cronholm et al. 1998).

The seed developmental stages are categorized similarly to those of rice. Grain development takes approximately 36 days before reaching maturity, passing through the flowering, milk, soft-dough, and hard-dough stages, which take approximately 8, 8, 10, and 10 days, respectively (e.g., Hall et al. 1983).

2.5   RICE

Rice is one of the three leading food crops in the world and provides 20% of the energy and 15% of the protein consumed by humans (Int. Rice Res. Inst. 1997a). Rice is grown on approximately 150 million ha in more than 50 countries in Asia, North and South America, Europe, Australia, and Africa (Int. Rice Res. Inst. 1997a, b). The United States is the second largest exporter of rice and accounts for 18% of internationally traded rice (Int. Rice Res. Inst. 1997b). Thus, the United States plays a major role in supplying rice to countries unable to meet domestic demand.

In 1997, rice was harvested from 3 million acres (1.2 million ha) in six states and was valued at approximately $1.7 billion (Table 2.1) (USDA Nat. Agric. Stat. 1998). Arkansas harvested the most acreage (1.37 million acres) followed by Louisiana, California, Texas, Mississippi, and Missouri (USDA Nat. Agric. Stat. 1998). Florida also harvests around 20,000 acres annually at a value of approximately $10 million (Schueneman 1993). Rice production in the United States is highly mechanized and technologically advanced, which partially explains the high yields compared with other rice-producing regions. Long-grain rice is grown predominantly in the South, whereas short and medium varieties are grown mainly in California (USDA Nat. Agric. Stat. 1998). Long-grain rice has a relatively high amylose content, which is associated with dry, fluffy cooking features unlike short- and medium-grain varieties, which are lower in amylose and stickier when cooked (Chang and Bardenas 1965).

Rice development is divided into three phases: (1) vegetative, from seed germination to panicle initiation; (2) reproductive, from panicle initiation to anthesis or flowering; and (3) ripening, from anthesis to maturity (Vergara 1991). In the United States, rice is direct-seeded into dry seedbeds using a tractor-pulled grain drill or an airplane, or seeded into flooded seedbeds using an airplane. Panicle initiation (flowering) generally occurs when the plant produces the maximum number of tillers and signals the start of the reproductive phase. When a bulge appears in the upper leaf sheath due to the expansion of the developing panicle, it is termed the boot stage. The ripening phase is from heading to maturity and is the growing period most susceptible to stink bug colonization and feeding. The early stage of grain development is termed the milk stage when the grain is filled with a milky liquid. The soft- and hard-dough stages are when the grain turns from liquid to a soft doughy consistency and then hardens. In temperate areas of rice production, this ripening lasts approximately 30 to 60 days (Vergara 1991), commonly around 30 days for cultivars adapted for the United States.

Time from seeding to harvest is generally between 110 and 150 days and is dependent on the cultivar, cultural practices, temperature, and photoperiod (Vergara 1991). In the South, most rice is flush-irrigated upon rice emergence through the soil. A flush is a flood that lasts approximately 1 to 2 days followed by a drain and is applied when soil moisture is low. Southern rice farmers typically flush rice for 3 to 6 weeks after emergence, then apply a permanent flood that is not removed until rice nears maturity. The rice paddies (flooded fields) are drained approximately 2 weeks prior to harvest to allow the soil to dry so the crop can be harvested. In California, most rice seed is soaked in water to initiate germination and sown by airplane into flooded paddies that remain flooded until rice nears maturity. In Florida, Louisiana, Mississippi, and Texas, where the growing season is longer than in the more northern rice-producing states, a significant percentage of rice is ratoon-cropped. The ratoon crop is the second crop that develops from the stubble of the main crop. The ratoon crop undergoes a ripening phase that is much later in the growing season, thus providing a suitable late-season host crop for stink bug colonization and development (M. O. Way, personal communication).

2.6 WHEAT

Wheat is the second leading food staple worldwide, behind only rice (Elliott et al. 1994). It is the most important small grain in the United States in total production, with approximately 2. 5 billion bu produced in 1997 (see Table 2.1) (USDA Nat. Agric. Stat. 1998).

In 1997, wheat was harvested from 63.6 million acres (27.7 million ha) in 42 states throughout all regions in the United States and was valued at $8.6 billion (Table 2.1) (USDA Nat. Agric. Stat. 1998). North Dakota (11.0 million acres) and Kansas (11.0 million acres) harvested the most acreage, followed by Montana (5.9 million acres), Oklahoma (5.4 million acres), and Texas (4.1 million acres); 14 states harvested 1 million acres or more (USDA Nat. Agric. Stat. 1998). The United States is the largest wheat-producing country in the world (USDA Nat. Agric. Stat. 1998).

Seed developmental stages are categorized similarly to those of rice. As with most crops, wheat is not susceptible to stink bug infestations and damage until after anthesis and grain development begin. Once anthesis is completed, the grain undergoes the milk stage during which time the grain is filled with a watery then milky substance. Then, the plant develops through the dough stages, first early dough, followed by soft dough, then hard dough when the liquid turns to a doughy consistency before hardening. Finally, the grain goes through the ripening stage as the grain seeds harden and the plant senesces (Ramseur et al. 1989).

2.7   ALFALFA

Alfalfa has been produced as a forage crop in the United States since about 1850 (Yeargan 1985). During the 1950s, a major expansion occurred in the north central states, while little change occurred elsewhere. Now, all regions of the country have large plantings of alfalfa except the southeastern states (Yeargan 1985).

In 1997, alfalfa was harvested from 23.7 million acres (9.6 million ha) nation-wide and was valued at $13.4 billion (Table 2.1) (USDA Nat. Agric. Stat. 1998). South Dakota harvested the most acreage (2.3 million acres), followed by Wisconsin (1.9 million acres), North Dakota (1.8 million acres), Montana (1.7 million acres), and Minnesota (1.5 million acres) (USDA Nat. Agric. Stat. 1998).

Forage alfalfa is a perennial crop with stands persisting for 3 to 7 years or longer. The crop may be planted as a pure stand or mixed with other forage grasses. Within the growing season, the crop is harvested two or three times in the more northern states and four or more times in the southern states (Yeargan 1985).

Alfalfa seed production is concentrated in the western United States (Yeargan 1985). In 1982, 88,522 lb (40,237 kg) were produced in the United States, 92.3% of which was produced in 11 western states. Most production was concentrated in California (43,911 lb), Idaho (10, 981 lb), and Washington (10,705 lb) (Rinckner 1991). Often, the first one or two alfalfa crops are cut for hay; then, the crop is allowed to produce seeds that are harvested later in the season (Russell 1952). Although stink bugs will feed readily on the alfalfa forage, because no fruiting structures are available, no serious crop losses are caused by these pests. Alfalfa produced for seed can have high populations of stink bugs and can sustain severe damage to immature seeds as a result of stink bug feeding. Rinckner (1991) mentioned the importance of insecticide selection for controlling alfalfa insect pests to minimize the detrimental impact on the pollinators that are essential for alfalfa seed production.

2.8   PECAN

Pecans are a native crop of North America and were noted by Spanish explorers in the early 1500s as hard-shelled nuts that grew along rivers and were consumed by Indians (Rice 1994). By the mid-1700s, pecans apparently had been distributed by American Indians throughout much of the southern United States and probably were planted both on purpose and by accident (Rice 1994).

In the early 1700s, there are references to pecan appearing in the French explorers' records; however, it is not known whether seeds of these nuts were sent to Europe (Crocker 1984a). In 1761, there is a possible record of the first shipment of pecans from America to London. Records from the late 1700s indicate pecans were sent from Louisiana to the North and also were planted in the colonies, where they were known as "Mississippi nuts" (Crocker 1984a).

A slave gardener named Antoine, who lived on a Louisiana plantation, appears to have been the first to graft pecan trees successfully. He grafted 16 trees from 1846 to 1847 and later worked on more (Crocker 1984a). By the end of the Civil War, there were 126 trees on the plantation of the variety named Centennial that were bearing nuts that buyers eagerly purchased. Thus, the foundation had been laid upon which a great industry was to develop (Crocker 1984a). About 1880, the commercial pecan industry was established when Indians and settlers began hauling pecans into San Antonio, TX, to sell (Brison 1986).

Since the introduction of pecan varieties, more than 200 different selections have been developed and sold. Many of these varieties have not survived, but a few are still available today. One of the named varieties that is grown commonly is the Stuart pecan, which was introduced in 1892 (Crocker 1984a). It was not until the early 1890s that grafted and budded trees of named varieties were available in sufficient numbers to plant in orchards, and southwest Georgia became one of the areas with the highest number of named varieties. Many of the early pecan groves were planted not to produce nuts but to increase the value of the real estate. Crocker (1984a) briefly reviewed the history of pecans in North America.

The vegetative and fruit growth periods of pecan follow an orderly sequence of events (Crocker 1984b). During each season, nut-bearing trees produce foliage, shoot growth, catkins (male flowers), and pistillate flowers (female flowers). This is followed by pollination and nut expansion. As the nuts grow, they form shucks, shells, and kernels. Mature pecan trees usually have a flush of growth immediately after bud break in the early spring. Rapid shoot elongation and leaf expansion occur during a 30-day period after bud break. During this initial flush of growth, catkins grow at the tip of 1-year-old wood where the current season's twigs originate. The catkin buds are produced in the spring, almost 1 year before they bloom (Crocker 1984b) and produce pollen to fertilize the female flowers. Toward the end of the rapid growth period, the pistillate flowers emerge on the end of the current season's growth and soon are ready for pollination (Crocker 1984b).

A successfully pollinated flower normally is indicated by a change of stigma color from yellowish to brown. Flowers that have not been pollinated and fertilized fall off the tree. Following the rapid growth period, leaf growth begins. After approximately one half of the leaf growth is completed, root growth may begin. When the leaf growth is approximately two thirds completed, secondary growth of branches, trunk, and roots occurs (Crocker 1984b).

Fruit growth usually lags until shoot and leaf growth have been completed. Then, the fruit undergoes a period of rapid expansion during which shuck and shell growth increase rapidly. After shell growth, rapid kernel growth and oil formation begin, followed by fruit maturity (Crocker 1984b).

Stink bugs can puncture the shuck and nut shell and extract material from the kernel (Rice 1994). Pecans are susceptible to attack during two long phenological stages: fruit enlargement and kernel development (Dutcher 1984). Before shell hardening, during the water stage (nut expansion and watery contents in the shell), the nuts bleed and abort (Rice 1994). After shell hardening and while the kernel is developing, stink bug feeding results in kernel spot (dark bitter spots on the damaged kernel). This damage reduces the quality of the nut and, in some cases, makes the crop unmarketable (Rice 1994).

In 1997, pecans were harvested in 14 states that produced 272 million pounds (123.6 million kg) valued at $257 million (Table 2.1) (USDA Nat. Agric. Stat. 1998). Georgia and Texas were the biggest producers and, combined, accounted for 57% of the nation's total production (USDA Nat. Agric. Stat. 1998).

2.9   MACADAMIA

Macadamia nuts represent an economically important crop in Hawaii. In 1989–1990, nuts were harvested from 21,900 acres (8,863 ha), totaling 47 million pounds (21.4 million kg) with a value of $43.2 million (Oi 1991). Similarly, in 1997, nuts were harvested from 19,200 acres (7,770 ha), totaling 58 million pounds (26.4 million kg) with a value of $42.9 million (Table 2.1) (USDA Nat. Agric. Stat. 1998).

2.10   FRUIT CROPS

In 1997, apples were produced in 36 states, with a total value of $1.7 billion (Table 2.1) (USDA Nat. Agric. Stat. 1998). Washington produced 47.9% of the national total, and New York, Michigan, and California, combined, produced an additional 30.8% (USDA Nat. Agric. Stat. 1998). Peaches were produced in 30 states, with a total value of $451.2 million (Table 2.1) (USDA Nat. Agric. Stat. 1998). California, Georgia, and South Carolina were the highest producing states, with 83.2% of the national total (USDA Nat. Agric. Stat. 1998). Pears were produced in nine states, with a total value of $299.6 million (Table 2.1) (USDA Nat. Agric. Stat. 1998). Washington, California, and Oregon produced 97.9% of the national total (USDA Nat. Agric. Stat. 1998).

2.11   COTTON

Cotton, a member of the family Malvaceae, is one of the most interesting higher organisms in the plant kingdom, producing both fiber and food. In its native state, cotton basically is a perennial woody shrub in a semidesert habitat (Brown 1986). Some of the other members in this diverse family include okra and different species of mallow that are actually weed pests in cotton (Tables 2.2 to 2.4).

In 1997, cotton was harvested from 13.3 million acres (5.4 million ha) in 17 contiguous states in the southern region of the United States from Virginia in the east to California in the west; the value was approximately $6.1 billion (Table 2.1) (USDA Nat. Agric. Stat. 1998). Texas, Georgia, California, Mississippi, and Arkansas harvested the most acreage, accounting for 72.1% of the national total (USDA Nat. Agric. Stat. 1998).

As with most other host plants, stink bugs are not a pest problem on cotton until the crop begins its flowering and boll development. The length of time between planting and the first square (bud) and flower depends on which node (usually the fifth to the tenth) the first fruiting branch appears. This requires approximately 60 days and is largely dependent on temperature (Mauney 1986). Once flowering begins, it continues over a long period of time as a function of vegetative growth that produces additional sites for fruiting branches. Flower buds or squares open as blossoms approximately 3 weeks after they first appear. A 1-day-old cotton bloom is white. On the second day, it turns red (or pink) (Ohlendorf 1996). After the first fruit or bolls are formed, the growth rate of the main cotton stem slows and eventually stops. Once all the floral bud sites found in this initial growth phase have matured into blossoms, there is a gap in flowering known as "cut-out" (Mauney 1986). Following cut-out, if sufficient moisture, nutrients, and heat units are available, there is resumption of vegetative growth followed by a new flush of flowering. Approximately 50 flower buds will blossom per plant over a period of approximately 60 days, and approximately 10 to 15 of these will mature into harvestable lint and seed (Mauney 1986).

Once pollinated, the seed begins to develop within the boll in the locks (the internal division of the boll into three to five seed chambers where the seed and fibers are produced) (Ohlendorf 1996). A single epidermal cell of the seed gives rise to a fiber (Brown 1986). Many fibers are produced by a single seed. Stink bugs feed on the developing seeds and associated fiber or lint (Barbour et al. 1988) and cause young bolls to fall off the plant, yellow staining of the lint, and reduced yields (Bundy et al. 1998). Stink bugs primarily damage young bolls during mid- to late season as adults migrate into fields from surrounding alternate host plants (Greene et al. 1998).

2.12   TOMATO

In 1997, fresh market tomatoes were harvested from approximately 125,000 acres (50,600 ha) in 20 states and were valued at approximately $1.2 billion (Table 2.1) (USDA Nat. Agric. Stat. 1998). California (40,800 acres) and Florida (38,100 acres) harvested the most acreage (USDA Nat. Agric. Stat. 1998). Processing tomatoes were harvested from 283,370 acres (114,725 ha) in 13 states and were valued at $605.4 million (Table 2.1) (USDA Nat. Agric. Stat. 1998). California (260,000 acres) harvested the most acreage. Stink bug feeding on both the green (immature) and the maturing fruits can cause quality and yield losses.

TABLE 2.2
Common Name, Scientific Name, and Family of Each Plant or Crop Mentioned Where Stink Bug Feeding and Development Were Observed or Implied, Alphabetized by Common Name[a]

Common Name	Scientific Name	Family[b]
Alfalfa	*Medicago sativa* L.	Leguminosae
Alfalfilla (see Indian sweet-clover)		
Alfileria	*Erodium cicutarium* (L.) L'Heritier	Geraniaceae
American dewberry	*Rubus flagellaris* Willdenow	Rosaceae
American holly	*Ilex opaca* Aiton	Aquifoliaceae
Annual blue grass	*Poa annua* L.	Gramineae
Apple	*Malus domestica* Borkhauser	Rosaceae
Apricot	*Prunus armeniaca* L.	Rosaceae
Arizona honeysweet	*Tidestroma oblongifolia* (S. Watson) Standley	Amaranthaceae
Asparagus	*Asparagus officinalis* L.	Liliaceae
Australian saltbush	*Atriplex semibaccata* R. Brown	Chenopodiaceae
Balsamroot	*Balsamorhiza sagittata* (Pursh) Nuttall	Compositae
Barley	*Hordeum vulgare* L.	Gramineae
Barnyard grass	*Echinochloa crusgalli* (L.) Beauvois	Gramineae
Basketflower	*Centaurea americana* Nuttall	Compositae
Bean (see snap bean)		
Beet	*Beta vulgaris* L.	Chenopodiaceae
Beggerweed	*Desmodium tortuosum* (Swartz) de Candolle	Leguminosae
Bell pepper	*Capsicum annuum* L.	Solanaceae
Bermuda grass	*Cynodon dactylon* (L.) Persoon	Gramineae
Bitterbrush	*Purshia tridentata* (Pursh) de Candolle	Rosaceae
Bitter cress	*Cardamine pratensis* L.	Cruciferae
Blackberry (see cut-leaf blackberry)		
Black cherry (see wild black cherry)		
Black locust	*Robinia pseudo-acacia* L.	Leguminosae
Black mustard	*Brassica nigra* (L.) Koch	Cruciferae
Bladderpod	*Isomeris arborea* Nuttall	Capparaceae
Bog marsh cress	*Rorippa islandica* (Oeder) Borbas	Cruciferae
Boxelder	*Acer negundo* L.	Aceraceae
Brittlebush	*Encelia farinosa* A. Gray	Compositae
Broad-leaved plantain	*Plantago major* L.	Plantaginaceae
Broccoli	*Brassica oleracea* L. var. *botrytis* L.	Cruciferae
Browntop panicum	*Panicum fasciculatum* Swartz	Gramineae
Browntop-millet	*Brachiaria fasciculata* (Swartz) Parodi	Gramineae
Brownweed	*Gutierrezia* spp.	Compositae
Buckwheat	*Fagopyrum esculentum* Moench	Polygonaceae
Bur-clover	*Medicago hispida* Gaertn	Leguminosae
Bur-marigold	*Bidens pilosa* L.	Compositae

TABLE 2.2 (CONTINUED)
Common Name, Scientific Name, and Family of Each Plant or Crop
Mentioned Where Stink Bug Feeding and Development Were Observed or
Implied, Alphabetized by Common Name[a]

Common Name	Scientific Name	Family[b]
Bur-sage	*Ambrosia acanthicarpa* Hooker [*Franseria acanthicarpa* (Hooker) Coville] *Ambrosia deltoidea* (Torrey) Payne	Compositae
Cabbage	*Brassica oleracea* L. var. *capitata* L.	Cruciferae
Camphorweed	*Heterotheca subaxillaris* (Lamarck) Britten & Rusby [*Heterotheca psammophila* Wagenknecht]	Compositae
Canaigre	*Rumex hymenosepalus* Torrey	Polygonaceae
Capper	*Cleome* spp.	Capparaceae
Carelessweed (Palmer amaranth)	*Amaranthus palmeri* S. Watson	Amaranthaceae
Cauliflower	*Brassica oleracea* L. var. *botrytis* L.	Cruciferae
Cheeseweed	*Malva parviflora* L.	Malvaceae
Cherry (sweet)	*Prunus avium* L.	Rosaceae
Chinese privet	*Ligustrum sinense* Loureiro	Oleaceae
Chinese pusley	*Heliotropium curassavicum* L.	Boraginaceae
Clammyweed	*Polanisia dodecandra* (L.) de Candolle	Cleomaceae
Clover	*Trifolium* spp.	Leguminosae
Coffee senna	*Senna occidentalis* (L.) Link [*Cassia occidentalis* L.]	Leguminosae
Collard	*Brassica oleraceae* L.	Cruciferae
Common cocklebur	*Xanthium strumarium* L.	Compositae
Common mullein (see woolly mullein)		
Common mustard	*Brassica kaber* (de Candolle) Wheeler	Cruciferae
Common saltbush (Desert saltbush)	*Atriplex polycarpa* (Torrey) Watson	Chenopodiaceae
Corn (field, sweet)	*Zea mays* L.	Gramineae
Cotton	*Gossypium hirsutum* L.	Malvaceae
Cowpea	*Vigna unguiculata* (L.) Walpers [*Vigna sinensis* (L.) Endlicher]	Leguminosae
Crab grass	*Digitaria sanguinalis* (L.) Scopoli	Gramineae
Creeping juniper	*Juniperus horizontalis* Moench	Cupressaceae
Creosote-bush	*Larrea tridentata* (Scsse & Mocino ex de Candolle) Coulter	Zygophyllaceae
Curly dock (see curlyleaf dock)		
Curlyleaf dock	*Rumex crispus* L.	Polygonaceae
Currant, Wax	*Ribes cereum* Douglas	Grossulariaceae
Cut-leaf blackberry	*Rubus laciniatus* Willdenow	Rosaceae
Dahlia	*Dahlia pinnata* Cavanilles	Compositae
Daisy fleabane	*Erigeron annuus* (L.) Person	Compositae
Dallis grass	*Paspalum dilatatum* Poiret	Gramineae
Desert sunflower	*Geraea canescens* Torrey & A. Gray	Compositae
Desert thorn	*Lycium pallidum* Miers	Solanaceae

TABLE 2.2 (CONTINUED)
Common Name, Scientific Name, and Family of Each Plant or Crop Mentioned Where Stink Bug Feeding and Development Were Observed or Implied, Alphabetized by Common Name[a]

Common Name	Scientific Name	Family[b]
Dogwood	*Cornus* spp.	Cornaceae
Dryas	*Dryas drummondii* Richardson	Rosaceae
Eastern manna-grass	*Glyceria septentrionalis* Hitchcock	Gramineae
Eggplant	*Solanum melongena* L.	Solanaceae
Elderberry	*Sambucus canadensis* L.	Caprifoliaceae
English wheat	*Triticum turgidum* L.	Gramineae
European strawberry	*Fragaria vesca* L.	Rosaceae
Fall panicum	*Panicum dichotomiflorum* Michaux	Gramineae
Fescue	*Festuca* spp.	Gramineae
Feterita (see grain sorghum)		
Field mustard (see turnip)		
Fig	*Ficus carica* L.	Moraceae
Flax	*Linum usitatissimum* L.	Linaceae
Fleabane	*Erigeron canadensis* L.	Compositae
Flowering dogwood	*Cornus florida* L.	Cornaceae
Fox grape (concord)	*Vitis labrusca* L.	Vitaceae
Foxtail millet (Foxtail grass)	*Setaria italica* (L.) Beauvois	Gramineae
Fragrant bitter weed	*Actinea odorata* Kuntze	Compositae
Gaping panicum	*Panicum hians* Elliott	Gramineae
Gaura	*Gaura parviflora* Douglas ex Lehmann	Onagraceae
Globe-mallow	*Sphaeralcea emoryi* Torrey ex A. Gray	Malvaceae
Grain sorghum	*Sorghum bicolor* (L.) Moench	Gramineae
Gramma	*Bouteloua* spp.	Gramineae
Grape	*Vitis* spp.	Vitaceae
Green bean (see snap bean)		
Hairy gaura (see woolly gaura)		
Hairy indigo	*Indigofera hirsuta* L.	Leguminosae
Honey locust	*Gleditsia triacanthos* L.	Leguminosae
Honeysuckle	*Lonicera involucrata* (Richardson) Banks	Caprifoliaceae
Hopsage (Spiny hopsage)	*Grayia spinosa* (Hooker) Moquin-Tandon	Chenopodiaceae
Horehound	*Marrubium vulgare* L.	Labiatae
Horse purslane	*Trianthema portulacastrum* L.	Aizoaceae
Horseweed	*Erigeron canadensis* L.	Compositae
Indian sweet-clover	*Melilotus indica* (L.) Allioni	Leguminosae
Iodine bush	*Allenrolfea occidentalis* (S. Watson) Kutze	Chenopodiaceae
Ironweed	*Vernonia altissima* Nuttall	Compositae
Italian ryegrass	*Lolium multiflorum* Lamarck	Gramineae
Jimsonweed	*Datura stramonium* L.	Solanaceae
Johnson grass	*Sorghum halepense* (L.) Persoon	Gramineae
Jojoba	*Simmondsia chinensis* (Link) Schneider	Buxaceae
Junglerice	*Echinochloa colona* (L.) Link	Gramineae
Juniper (see mountain juniper)		

TABLE 2.2 (CONTINUED)
Common Name, Scientific Name, and Family of Each Plant or Crop
Mentioned Where Stink Bug Feeding and Development Were Observed or
Implied, Alphabetized by Common Name[a]

Common Name	Scientific Name	Family[b]
Kafir corn (Broom corn)	*Sorghum vulgare* Persoon	Gramineae
Kale	*Brassica oleracea* L. var. *acephala* de Candolle	Cruciferae
Klein grass	*Panicum coloratum* L.	Gramineae
Knot grass	*Paspalum distichum* L.	Gramineae
Kohlrabi	*Brassica oleracea* L. var. *gongylodes* L.	Cruciferae
Lamb's quarters	*Chenopodium album* L.	Chenopodiaceae
Lance-leaf crotalaria	*Crotalaria lanceolata* E. Meyer	Leguminosae
Lentil	*Lens culinaris* Medicus	Leguminosae
Lettuce	*Lactuca sativa* L.	Compositae
Lima bean	*Phaseolus lunatus* L.	Leguminosae
Linden	*Tilia* spp.	Tiliaceae
Linseed	*Linum usitatissimum* L.	Linaceae
Little-seed canary grass	*Phalaris minor* Retzius	Gramineae
Lizard's tail (see velvetweed)		
Lodgepole pine	*Pinus contorta* Douglas	Pinaceae
London rocket	*Sisymbrium irio* L.	Cruciferae
Macadamia	*Macadamia integrifolia* Maiden & Betche	Proteaceae
Mallow	*Malva parviflora* L.	Malvaceae
May-pop	*Passiflora incarnata* L.	Passifloraceae
Mesquite	*Prosopis juliflora* (Swartz) de Candolle	Leguminosae
Milo (see grain sorghum)		
Milo maize (see grain sorghum)		
Mimosa	*Albizia julibrissin* Durazzini	Leguminosae
Mormon tea	*Ephedra aspera* Engelmann	Ephedraceae
	Ephedra spp.	
Mountain juniper	*Juniperus communis* L.	Cupressaceae
Mung bean	*Vigna radiata* (L.) Wilczek	Leguminosae
Nettle leaf goose foot	*Chenopodium murale* L.	Chenopodiaceae
Nodding beak rush	*Rhynchospora inexpansa* (Michaux) Vahl	Cyperaceae
Nutsedge	*Cyperus esculentus* L.	Cyperaceae
Oatgrass	*Stipa avenacea* L.	Gramineae
Oats	*Avena sativa* L.	Gramineae
Okra	*Abelmoschus esculentus* (L.) Moench	Malvaceae
Orange	*Citrus sinensis* (L.) Osbeck	Rutaceae
Partridge pea	*Chamaecrista fasciculata* (Michaux) Greene [*Cassia fasciculata* Michaux]	Leguminosae
Paspalum	*Paspalum longipilum* Nash [*Paspalum pubiflorum* Ruprecht]	Gramineae
Pea (field, English)	*Pisum sativum* L.	Leguminosae
Peach	*Prunus persica* (L.) Batsch	Rosaceae
Peanut	*Arachis hypogaea* L.	Leguminosae
Pear	*Pyrus communis* L.	Rosaceae

TABLE 2.2 (CONTINUED)
Common Name, Scientific Name, and Family of Each Plant or Crop
Mentioned Where Stink Bug Feeding and Development Were Observed or
Implied, Alphabetized by Common Name[a]

Common Name	Scientific Name	Family[b]
Pecan	*Carya illinoensis* (Wangenheim) K. Koch	Juglandaceae
Pepper (see bell pepper)		
Peppercress (see peppergrass)		
Peppergrass	*Lepidium virginicum* L.	Cruciferae
Peppervine	*Ampelopsis arborea* (L.) Koehne	Vitaceae
Perennial ryegrass	*Lolium perenne* L.	Gramineae
Photinia	*Photinia serrulata* L.	Rosaceae
Pigweed	*Amaranthus hybridus* L.	Amaranthaceae
Pistachio	*Pistacia vera* L.	Anacardiaceae
Plum	*Prunus domestica* L.	Rosaceae
Pointleaf manzanita	*Arctostaphylos pungens* Kunth	Ericaceae
Ponderosa pine	*Pinus ponderosa* Douglas	Pinaceae
Potato	*Solanum tuberosum* L.	Solanaceae
Prairie bean	*Sesbania emerus* (Aublet) Urban	Leguminosae
Prickly lettuce	*Lactuca serriola* L.	Compositae
	[*Lactuca scariola* L.]	
Prickly pear	*Opuntia humifusa* Rafinesque	Cactaceae
Purple-flowered thistle	*Cirsium virginianum* L.	Compositae
Radish	*Raphanus sativus* L.	Cruciferae
Rape	*Brassica napus* L.	Cruciferae
Raspberry (see red raspberry)		
Redbud	*Cercis* spp.	Leguminosae
Red raspberry	*Rubus idaeus* L.	Rosaceae
Rice	*Oryza sativa* L.	Gramineae
Rough-leaf dogwood	*Cornus drummondi* C.A. Meyer	Cornaceae
Russian thistle	*Salsola iberica* Sennen & Pau	Chenopodiaceae
	Salsola pestifer Nelson	
Rutabaga	*Brassica napus* L.	Cruciferae
Rye	*Secale cereale* L.	Gramineae
Sage	*Salvia officinalis* L.	Labiatae
Saltbush (Fourwing)	*Atriplex canescens* (Pursh) Nuttall	Chenopodiaceae
Scots' pine (Scotch pine)	*Pinus sylvestris* L.	Pinaceae
Sedge	*Carex* spp.	Cyperaceae
Sesame	*Sesamum indicum* L.	Pedaliaceae
Sheepweed	*Gutierrezia* spp.	Compositae
Shepherd's purse	*Capsella bursa-pastoris* (L.) Medicus	Cruciferae
Showy crotalaria	*Crotalaria spectabilis* Roth	Leguminosae
Silvery saltbush	*Atriplex argentea* Nuttall	Chenopodiaceae
Skunk bush (Squaw bush)	*Rhus trilobata* Torrey & A. Gray	Anacardiaceae
Slender-leaf crotalaria	*Crotalaria brevidens* Bentham	Leguminosae
Slimpod senna	*Cassia leptocarpa* Bentham	Leguminosae

TABLE 2.2 (CONTINUED)
Common Name, Scientific Name, and Family of Each Plant or Crop Mentioned Where Stink Bug Feeding and Development Were Observed or Implied, Alphabetized by Common Name[a]

Common Name	Scientific Name	Family[b]
Snakeweed	*Gutierrezia* spp.	Compositae
Snap bean	*Phaseolus vulgaris* L.	Leguminosae
Sorghum (see grain sorghum)		
Sour-clover (see Indian sweet-clover)		
Southern pea (see cowpea)		
Sowthistle	*Sonchus oleraceous* L.	Compositae
Soybean	*Glycine max* (L.) Merrill	Leguminosae
Spanish moss	*Tillandsia usneoides* L.	Bromeliaceae
Spelt (see wheat)		
Spiderwort	*Tradescantia* spp.	Commelinaceae
Spinach	*Spinacia oleracea* L.	Chenopodiaceae
Spruce	*Picea* spp.	Pinaceae
Squash	*Cucurbita maxima* Duchesne	Cucurbitaceae
Strawberry	*Fragaria* spp.	Rosaceae
Sudan grass	*Sorghum sudanese* (Piper) Stapf	Gramineae
Sugar beet (see beet)		
Sugarcane	*Saccharum officinarum* L.	Gramineae
Sumac (see winged sumac)		
Summer cypress	*Kochia scoparia* (L.) Schrad	Chenopodiaceae
Sunflower	*Helianthus annuus* L.	Compositae
Sweet grass	*Glyceria septentrionalis* Hitchcock	Gramineae
Tahoka daisy	*Machaeranthera tanacetifolia* (Humboldt, Bonpland, Kunth) Nees	Compositae
Tangerine	*Citrus reticulata* Blanco	Rutaceae
Tansy mustard	*Descurainia pinnata* (Walter) Britton	Cruciferae
Texas millet	*Brachiaria texana* Buckley	Gramineae
Texas thistle	*Cirsium texanum* Buckley	Compositae
Tobacco	*Nicotiana tabacum* L.	Solanaceae
Tomato	*Lycopersicon esculentum* Miller	Solanaceae
Trumpet-creeper	*Campsis radicans* (L.) Seeman	Bignoniaceae
Tumble mustard	*Sisymbrium altissimum* L.	Cruciferae
Tumbleweed	*Salsola kali* L. var. *tenuifolia* Tausch	Chenopodiaceae
Twinberry	*Lonicera involucrata* (Richardson) Banks	Caprifoliaceae
Turnip	*Brassica rapa* L.	Cruciferae
Vasey grass	*Paspalum urvillei* Steudel	Gramineae
Velvet bean	*Stizolobium deeringianum* Bort	Leguminosae
Velvet mesquite	*Prosopis velutina* Wooton	Leguminosae
Velvetweed	*Gaura parviflora* Douglas ex Lehmann	Onagraceae
Vetch	*Vicia angustifolia* Reichard	Leguminosae
Viper's bugloss	*Echium vulgare* L.	Boraginaceae
Western dock	*Rumex occidentalis* S. Watson	Polygonaceae

TABLE 2.2 (CONTINUED)
Common Name, Scientific Name, and Family of Each Plant or Crop
Mentioned Where Stink Bug Feeding and Development Were Observed or
Implied, Alphabetized by Common Name[a]

Common Name	Scientific Name	Family[b]
Wheat	*Triticum aestivum* L.	Gramineae
	Triticum vulgare Villars	
White campion	*Lychnis alba* Miller	Caryophyllaceae
White horse-nettle	*Solanum elaeagnifolium* Cavanilles	Solanaceae
White-top fleabane	*Erigeron annuus* (L.) Persoon	Compositae
Wild black cherry	*Prunus serotina* Ehrhart	Rosaceae
Wild buckwheat	*Eriogonum* spp.	Polygonaceae
Wild grape	*Vitis* spp.	Vitaceae
Wild heliotrope	*Heliotropium curassavicum* L.	Boraginaceae
Wild lettuce	*Lactuca canadensis* L.	Compositae
Wild oats-slender	*Avena barbata* Brotero	Gramineae
Wild oats-tartarian	*Avena fatua* L.	Gramineae
Wild radish	*Raphanus raphanistrum* L.	Cruciferae
Winged sumac	*Rhus copallina* L.	Anacardiaceae
Winter emmer	*Triticum turgidum* L.	Gramineae
Winter rape (see rape)		
Woolly gaura	*Gaura villosa* Torrey	Onagraceae
Woolly mullein	*Verbascum thapsus* L.	Scrophulariaceae
Wormwood	*Artemisia* spp.	Compositae
Yellow mustard	*Brassica* spp.	Cruciferae
Yucca	*Yucca aloifolia* L.	Agavaceae

[a] All scientific names were verified with one or more authoritative references including Fernald 1950, Bailey and Bailey 1976, Radford et al. 1981, Faust and Strang 1983, Weed Sci. Soc. Am. 1984, and Brako et al. 1997.

[b] Depending on taxonomic source, some plant family names are different than those appearing in this table. These names include Lamiaceae (= Labiatae), Fabaceae (= Leguminosae), Poaceae (= Gramineae), Asteraceae (= Compositae), Brassicaceae (= Cruciferae), and Simmondiaceae (= Buxaceae) (Brako et al. 1997). Although five of these names do not end in -aceae, they have been sanctioned by the International Code of Botanical Nomenclature because of old, traditional usage (Jones and Luchsinger 1986).

TABLE 2.3
Scientific Name, Common Name, and Family of Each Plant or Crop Mentioned Where Stink Bug Feeding and Development Were Observed or Implied, Alphabetized by Scientific Name[a]

Scientific Name	Common Name	Family[b]
Abelmoschus esculentus (L.) Moench	Okra	Malvaceae
Acer negundo L.	Boxelder	Aceraceae
Actinea odorata Kuntze	Fragrant bitter weed	Compositae
Albizia julibrissin Durazzini	Mimosa	Leguminosae
Allenrolfea occidentalis (S. Watson) Kutze	Iodine bush	Chenopodiaceae
Allium cepa L.	Onion	Liliaceae
Amaranthus hybridus L.	Pigweed	Amaranthaceae
Amaranthus palmeri S. Watson	Carelessweed, Palmer amaranth	Amaranthaceae
Ambrosia acanthicarpa Hooker [*Franseria acanthicarpa* (Hooker) Coville]	Bur-sage	Compositae
Ambrosia deltoidea (Torrey) Payne	Bur-sage	Compositae
Ampelopsis arborea (L.) Koehne	Peppervine	Vitaceae
Arachis hypogaea L.	Peanut	Leguminosae
Arctostaphylos pungens Kunth	Pointleaf manzanita	Ericaceae
Artemisia spp.	Wormwood	Compositae
Asparagus officinalis L.	Asparagus	Liliaceae
Atriplex argentea Nuttall	Silvery saltbush	Chenopodiaceae
Atriplex canescens (Pursh) Nuttall	Saltbush (Fourwing)	Chenopodiaceae
Atriplex polycarpa (Torrey) Watson	Common saltbush, Desert saltbush	Chenopodiaceae
Atriplex semibaccata R. Brown	Australian saltbush	Chenopodiaceae
Avena barbata Brotero	Slender wild oats	Gramineae
Avena fatua L.	Tartarian wild oats	Gramineae
Avena sativa L.	Oats	Gramineae
Balsamorhiza sagittata (Pursh) Nuttall	Balsamroot	Compositae
Beta vulgaris L.	Beet, Sugar beet	Chenopodiaceae
Bidens pilosa L.	Bur-marigold	Compositae
Bouteloua spp.	Gramma	Gramineae
Brachiaria fasciculata (Swartz) Parodi	Browntop-millet	Gramineae
Brachiaria texana Buckley	Texas millet	Gramineae
Brassica kaber (de Candolle) Wheeler	Common mustard	Cruciferae
Brassica napus L.	Rape, Rutabaga, Winter rape	Cruciferae
Brassica nigra (L.) Koch	Black mustard	Cruciferae
Brassica oleracea L.	Cabbage, Kale, Kholrabi, Collard	Cruciferae
Brassica oleracea L. var. *acephala* de Candolle	Kale	Cruciferae
Brassica oleracea L. var. *botrytis* L.	Broccoli, Cauliflower	Cruciferae
Brassica oleracea L. var. *capitata* L.	Cabbage	Cruciferae
Brassica oleracea L. var. *gongylodes* L.	Kohlrabi	Cruciferae
Brassica rapa L.	Turnip, Field mustard	Cruciferae
Brassica spp.	Yellow mustard	Cruciferae

TABLE 2.3 (CONTINUED)
Scientific Name, Common Name, and Family of Each Plant or Crop Mentioned Where Stink Bug Feeding and Development Were Observed or Implied, Alphabetized by Scientific Name[a]

Scientific Name	Common Name	Family[b]
Campsis radicans (L.) Seeman	Trumpet-creeper	Bignoniaceae
Capsella bursa-pastoris (L.) Medicus	Shepherd's purse	Cruciferae
Capsicum annuum L.	Bell pepper, Pepper	Solanaceae
Cardamine pratensis L.	Bitter cress	Cruciferae
Carex spp.	Sedge	Cyperaceae
Carya illinoensis (Wangenheim) K. Koch	Pecan	Juglandaceae
Cassia leptocarpa Bentham	Slimpod senna	Leguminosae
Centaurea americana Nuttall	Basketflower	Compositae
Cercis spp.	Redbud	Leguminosae
Chamaecrista fasciculata (Michaux) Greene [*Cassia fasciculata* Michaux]	Partridge pea	Leguminosae
Chenopodium album L.	Lamb's quarters	Chenopodiaceae
Chenopodium murale L.	Nettle leaf goose foot	Chenopodiaceae
Cirsium texanum Buckley	Texas thistle	Compositae
Cirsium virginianum L.	Purple-flowered thistle	Compositae
Citrus reticulata Blanco	Tangerine	Rutaceae
Citrus sinensis (L.) Osbeck	Orange	Rutaceae
Cleome spp.	Capper	Capparaceae
Cornus drummondi C.A. Meyer	Rough-leaf dogwood	Cornaceae
Cornus florida L.	Flowering dogwood	Cornaceae
Cornus spp.	Dogwood	Cornaceae
Crotalaria brevidens Bentham	Slender-leaf crotalaria	Leguminosae
Crotalaria lanceolata E. Meyer	Lance-leaf crotalaria	Leguminosae
Crotalaria spectabilis Roth	Showy crotalaria	Leguminosae
Cucurbita maxima Duchesne	Squash	Cucurbitaceae
Cynodon dactylon (L.) Persoon	Bermuda grass	Gramineae
Cyperus esculentus L.	Nutsedge	Cyperaceae
Dahlia pinnata Cavanilles	Dahlia	Compositae
Datura stramonium L.	Jimsonweed	Solanaceae
Descurainia pinnata (Walter) Britton	Tansy mustard	Cruciferae
Desmodium tortuosum (Swartz) de Candolle	Beggerweed	Leguminosae
Digitaria sanguinalis (L.) Scopoli	Crab grass	Gramineae
Dryas drummondii Richardson	Dryas	Rosaceae
Echinochloa colona (L.) Link	Junglerice	Gramineae
Echinochloa crusgalli (L.) Beauvois	Barnyard grass	Gramineae
Echium vulgare L.	Viper's bugloss	Boraginaceae
Encelia farinosa A. Gray	Brittlebush	Compositae
Ephedra aspera Engelmann	Mormon tea	Ephedraceae
Ephedra spp.	Mormon tea	Ephedraceae
Erigeron annuus (L.) Persoon	Daisy fleabane, White-top fleabane	Compositae
Erigeron canadensis L.	Horseweed, Fleabane	Compositae

TABLE 2.3 (CONTINUED)
Scientific Name, Common Name, and Family of Each Plant or Crop Mentioned Where Stink Bug Feeding and Development Were Observed or Implied, Alphabetized by Scientific Name[a]

Scientific Name	Common Name	Family[b]
Eriogonum spp.	Wild buckwheat	Polygonaceae
Erodium cicutarium (L.) L'Heritier	Alfileria	Geraniaceae
Fagopyrum esculentum Moench	Buckwheat	Polygonaceae
Festuca spp.	Fescue	Gramineae
Ficus carica L..	Fig	Moraceae
Fragaria vesca L.	European strawberry	Rosaceae
Fragaria spp.	Strawberry	Rosaceae
Gaura parviflora Douglas ex Lehmann	Gaura, Lizard's tail, Velvetweed	Onagraceae
Gaura villosa Torrey	Woolly gaura, Hairy gaura	Onagraceae
Geraea canescens Torrey & A. Gray	Desert sunflower	Compositae
Gleditsia triacanthos L.	Honey locust	Leguminosae
Glyceria septentrionalis Hitchcock	Eastern manna-grass, Sweet grass	Gramineae
Glycine max (L.) Merrill	Soybean	Leguminosae
Gossypium hirsutum L.	Cotton	Malvaceae
Grayia spinosa (Hooker) Moquin-Tandon	Hopsage (Spiny hopsage)	Chenopodiaceae
Gutierrezia spp.	Sheepweed, Snakeweed, Brownweed	Compositae
Helianthus annuus L.	Sunflower	Compositae
Heliotropium curassavicum L.	Wild heliotrope, Chinese pusley	Boraginaceae
Heterotheca subaxillaris (Lamarck) Britten & Rusby [*Heterotheca psammophila* Wagenknecht]	Camphorweed	Compositae
Hordeum vulgare L.	Barley	Gramineae
Ilex opaca Aiton	American holly	Aquifoliaceae
Indigofera hirsuta L.	Hairy indigo	Leguminosae
Isomeris arborea Nuttall	Bladderpod	Capparaceae
Juniperus communis L.	Mountain juniper, Juniper	Cupressaceae
Juniperus horizontalis Moench	Creeping juniper	Cupressaceae
Kochia scoparia (L.) Schrad	Summer cypress	Chenopodiaceae
Lactuca canadensis L.	Wild lettuce	Compositae
Lactuca sativa L.	Lettuce	Compositae
Lactuca serriola L. [*Lactuca scariola* L.]	Prickly lettuce	Compositae
Larrea tridentata (Sesse & Mocino ex de Candolle) Coulter	Creosote-bush	Zygophyllaceae
Lens culinaris Medicus	Lentil	Leguminosae
Lepidium virginicum L.	Peppergrass, Peppercress	Cruciferae
Ligustrum sinense Loureiro	Chinese privet	Oleaceae
Linum usitatissimum L.	Flax, Linseed	Linaceae
Lolium multiflorum Lamarck	Italian ryegrass	Gramineae
Lolium perenne L.	Perennial ryegrass	Gramineae

TABLE 2.3 (CONTINUED)
Scientific Name, Common Name, and Family of Each Plant or Crop
Mentioned Where Stink Bug Feeding and Development Were Observed or
Implied, Alphabetized by Scientific Name[a]

Scientific Name	Common Name	Family[b]
Lonicera involucrata (Richardson) Banks	Honeysuckle, Twinberry	Caprifoliaceae
Lychnis alba Miller	White campion	Caryophyllaceae
Lycium pallidum Miers	Desert thorn	Solanaceae
Lycopersicon esculentum Miller	Tomato	Solanaceae
Macadamia integrifolia Maiden & Betche	Macadamia	Proteaceae
Machaeranthera tanacetifolia (Humboldt, Bonpland, Kunth) Nees	Tahoka daisy	Compositae
Malus domestica Borkhauser	Apple	Rosaceae
Malva parviflora L.	Mallow, Cheeseweed	Malvaceae
Marrubium vulgare L.	Horehound	Labiatae
Medicago hispida Gaertn	Bur-clover	Leguminosae
Medicago sativa L.	Alfalfa	Leguminosae
Melilotus indica (L.) Allioni	Indian sweet-clover, Alfalfilla, Sour-clover	Leguminosae
Nicotiana tabacum L.	Tobacco	Solanaceae
Opuntia humifusa Rafinesque	Prickly pear	Cactaceae
Oryza sativa L.	Rice	Gramineae
Panicum coloratum L.	Klein grass	Gramineae
Panicum dichotomiflorum Michaux	Fall panicum	Gramineae
Panicum fasciculatum Swartz	Browntop panicum	Gramineae
Panicum hians Elliott	Gaping panicum	Gramineae
Paspalum dilatatum Poiret	Dallis grass	Gramineae
Paspalum distichum L.	Knot grass	Gramineae
Paspalum longipilum Nash	Paspalum	Gramineae
Paspalum pubiflorum Ruprecht	Paspalum	Gramineae
Paspalum urvillei Steudel	Vasey grass	Gramineae
Passiflora incarnata L.	May-pop	Passifloraceae
Phalaris minor Retzius	Little-seed canary grass	Gramineae
Phaseolus lunatus L.	Lima bean	Leguminosae
Phaseolus vulgaris L.	Snap bean, Bean, Green bean	Leguminosae
Photinia serrulata L.	Photinia	Rosaceae
Picea spp.	Spruce	Pinaceae
Pinus contorta Douglas	Lodgepole pine	Pinaceae
Pinus ponderosa Douglas	Ponderosa pine	Pinaceae
Pinus sylvestris L.	Scots' pine (Scotch pine)	Pinaceae
Pistacia vera L.	Pistachio	Anacardiaceae
Pisum sativum L.	Pea (field, English)	Leguminosae
Plantago major L.	Broad-leaved plantain	Plantaginaceae
Poa annua L.	Annual blue grass	Gramineae
Polanisia dodecandra (L.) de Candolle	Clammyweed	Cleomaceae
Prosopis juliflora (Swartz) de Candolle	Mesquite	Leguminosae
Prosopis velutina Wooton	Velvet mesquite	Leguminosae

TABLE 2.3 (CONTINUED)
Scientific Name, Common Name, and Family of Each Plant or Crop Mentioned Where Stink Bug Feeding and Development Were Observed or Implied, Alphabetized by Scientific Name[a]

Scientific Name	Common Name	Family[b]
Prunus armeniaca L.	Apricot	Rosaceae
Prunus avium L.	Cherry (sweet)	Rosaceae
Prunus domestica L.	Plum	Rosaceae
Prunus persica (L.) Batsch	Peach	Rosaceae
Prunus serotina Ehrhart	Wild black cherry, Black cherry	Rosaceae
Purshia tridentata (Pursh) de Candolle	Bitterbrush	Rosaceae
Pyrus communis L.	Pear	Rosaceae
Raphanus raphanistrum L.	Wild radish	Cruciferae
Raphanus sativus L.	Radish	Cruciferae
Rhus copallina L.	Winged sumac, Sumac	Anacardiaceae
Rhus trilobata Torrey & A. Gray	Skunk bush, Squaw bush	Anacardiaceae
Rhynchospora inexpansa (Michaux) Vahl	Nodding beak rush	Cyperaceae
Ribes cereum Douglas	Currant, Wax	Grossulariaceae
Robinia pseudo-acacia L.	Black locust	Leguminosae
Rorippa islandica (Oeder) Borbas	Bog marsh cress	Cruciferae
Rubus flagellaris Willdenow	American dewberry	Rosaceae
Rubus idaeus L.	Red raspberry, Raspberry	Rosaceae
Rubus laciniatus Willdenow	Cut-leaf blackberry, Blackberry	Rosaceae
Rumex crispus L.	Curlyleaf dock, Curly dock	Polygonaceae
Rumex hymenosepalus Torrey	Canaigre	Polygonaceae
Rumex occidentalis S. Watson	Western dock	Polygonaceae
Saccharum officinarum L.	Sugarcane	Gramineae
Salsola iberica Sennen & Pau	Russian thistle	Chenopodiaceae
Salsola kali L. var. *tenuifolia* Tausch	Tumbleweed	Chenopodiaceae
Salsola pestifer Nelson	Russian thistle	Chenopodiaceae
Salvia officinalis L.	Sage	Labiatae
Sambucus canadensis L.	Elderberry	Caprifoliaceae
Secale cereale L.	Rye	Gramineae
Senna occidentalis (L.) Link [*Cassia occidentalis* L.]	Coffee senna	Leguminosae
Sesamum indicum L.	Sesame	Pedaliaceae
Sesbania emerus (Aublet) Urban	Prairie bean	Leguminosae
Setaria italica (L.) Beauvois	Foxtail millet (Foxtail grass)	Gramineae
Simmondsia chinensis (Link) Schneider	Jojoba	Buxaceae
Sisymbrium altissimum L.	Tumble mustard	Cruciferae
Sisymbrium irio L.	London rocket	Cruciferae
Solanum elaeagnifolium Cavanilles	White horse-nettle	Solanaceae
Solanum melongena L.	Eggplant	Solanaceae
Solanum tuberosum L.	Potato	Solanaceae
Sonchus oleraceus L.	Sowthistle	Compositae
Sorghum bicolor (L.) Moench	Grain sorghum, Sorghum, Milo, Feterita, Milo maize	Gramineae

TABLE 2.3 (CONTINUED)
Scientific Name, Common Name, and Family of Each Plant or Crop Mentioned Where Stink Bug Feeding and Development Were Observed or Implied, Alphabetized by Scientific Name[a]

Scientific Name	Common Name	Family[b]
Sorghum halepense (L.) Persoon	Johnson grass	Gramineae
Sorghum sudanese (Piper) Stapf	Sudan grass	Gramineae
Sorghum vulgare Persoon	Kafir corn (Broom corn)	Gramineae
Sphaeralcea emoryi Torrey ex A. Gray	Globe-mallow	Malvaceae
Spinacia oleracea L.	Spinach	Chenopodiaceae
Stipa avenacea L.	Oatgrass	Gramineae
Stizolobium deeringianum Bort	Velvet bean	Leguminosae
Tidestroma oblongifolia (S. Watson) Standley	Arizona honeysweet	Amaranthaceae
Tilia spp.	Linden	Tiliaceae
Tillandsia usneoides L.	Spanish moss	Bromeliaceae
Tradescantia spp.	Spiderwort	Commelinaceae
Trianthema portulacastrum L.	Horse purslane	Aizoaceae
Trifolium spp.	Clover	Leguminosae
Triticum aestivum L.	Wheat (Spelt)	Gramineae
Triticum turgidum L.	English wheat, Winter emmer	Gramineae
Triticum vulgare Villars	Wheat	Gramineae
Verbascum thapsus L.	Woolly mullein, Common mullein	Scrophulariaceae
Vernonia altissima Nuttall	Ironweed	Compositae
Vicia angustifolia Reichard	Vetch	Leguminosae
Vigna radiata (L.) Wilczek	Mung bean	Leguminosae
Vigna unguiculata (L.) Walpers [*Vigna sinensis* (L.) Endlicher]	Cowpea, Southern pea	Leguminosae
Vitis labrusca L.	Fox grape (concord)	Vitaceae
Vitis spp.	Grape, Wild grape	Vitaceae
Xanthium strumarium L.	Common cocklebur	Compositae
Yucca aloifolia L.	Yucca	Agavaceae
Zea mays L.	Corn (field, sweet)	Gramineae

[a] All scientific names were verified with one or more authoritative references including Fernald 1950, Bailey and Bailey 1976, Radford et al. 1981, Faust and Strang 1983, Weed Sci. Soc. Am. 1984, and Brako et al. 1997.

[b] Depending on taxonomic source, some plant family names are different than those appearing in this table. These names include Lamiaceae (= Labiatae), Fabaceae (= Leguminosae), Poaceae (= Gramineae), Asteraceae (= Compositae), Brassicaceae (= Cruciferae), and Simmondiaceae (= Buxaceae) (Brako et al. 1997). Although five of these names do not end in -aceae, they have been sanctioned by the International Code of Botanical Nomenclature because of old, traditional usage (Jones and Luchsinger 1986).

TABLE 2.4
Family, Scientific Name, and Common Name of Each Plant or Crop Mentioned Where Stink Bug Feeding and Development Were Observed or Implied, Alphabetized by Family[a]

Family[b]	Scientific Name	Common name
Aceraceae	*Acer negundo* L.	Boxelder
Agavaceae	*Yucca aloifolia* L.	Yucca
Aizoaceae	*Trianthema portulacastrum* L.	Horse purslane
Amaranthaceae	*Tidestroma oblongifolia* (S. Watson) Standley	Arizona honeysweet
Amaranthaceae	*Amaranthus hybridus* L.	Pigweed
Amaranthaceae	*Amaranthus palmeri* S. Watson	Carelessweed (Palmer amaranth)
Anacardiaceae	*Pistacia vera* L.	Pistachio
Anacardiaceae	*Rhus copallina* L.	Winged sumac
Anacardiaceae	*Rhus trilobata* Torrey & A. Gray	Skunk bush, Squaw bush
Aquifoliaceae	*Ilex opaca* Aiton	American holly
Bignoniaceae	*Campsis radicans* (L.) Seeman	Trumpet-creeper
Boraginaceae	*Echium vulgare* L.	Viper's bugloss
Boraginaceae	*Heliotropium curassavicum* L.	Wild heliotrope, Chinese pusley
Bromeliaceae	*Tillandsia usneoides* L.	Spanish moss
Buxaceae	*Simmondsia chinensis* (Link) Schneider	Jojoba
Cactaceae	*Opuntia humifusa* Rafinesque	Prickly pear
Capparaceae	*Cleome* spp.	Capper
Capparaceae	*Isomeris arborea* Nuttall	Bladderpod
Caprifoliaceae	*Lonicera involucrata* (Richardson) Banks	Honeysuckle, Twinberry
Caprifoliaceae	*Sambucus canadensis* L.	Elderberry
Caryophyllaceae	*Lychnis alba* Miller	White campion
Chenopodiaceae	*Allenrolfea occidentalis* (S. Watson) Kutze	Iodine bush
Chenopodiaceae	*Atriplex argentea* Nuttall	Silvery saltbush
Chenopodiaceae	*Atriplex canescens* (Pursh) Nuttall	Saltbush (Fourwing)
Chenopodiaceae	*Atriplex polycarpa* (Torrey) Watson	Common saltbush, Desert saltbush
Chenopodiaceae	*Atriplex semibaccata* R. Brown	Australian saltbush
Chenopodiaceae	*Beta vulgaris* L.	Beet, Sugar beet
Chenopodiaceae	*Chenopodium album* L.	Lamb's quarters
Chenopodiaceae	*Chenopodium murale* L.	Nettle leaf goose foot
Chenopodiaceae	*Grayia spinosa* (Hooker) Moquin-Tandon	Hopsage, Spiny hopsage
Chenopodiaceae	*Kochia scoparia* (L.) Schrad	Summer cypress
Chenopodiaceae	*Spinacia oleracea* L.	Spinach
Chenopodiaceae	*Salsola iberica* Sennen & Pau	Russian thistle
Chenopodiaceae	*Salsola kali* L. var. *tenuifolia* Tausch	Tumbleweed
Chenopodiaceae	*Salsola pestifer* Nelson	Russian thistle
Cleomaceae	*Polanisia dodecandra* (L.) de Candolle	Clammyweed
Commelinaceae	*Tradescantia* spp.	Spiderwort
Compositae	*Actinea odorata* Kuntze	Fragrant bitter weed
Compositae	*Ambrosia acanthicarpa* Hooker [*Franseria acanthicarpa* (Hooker) Coville]	Bur-sage
Compositae	*Ambrosia deltoidea* (Torrey) Payne	Bur-sage

TABLE 2.4 (CONTINUED)
Family, Scientific Name, and Common Name of Each Plant or Crop Mentioned Where Stink Bug Feeding and Development Were Observed or Implied, Alphabetized by Family[a]

Family[b]	Scientific Name	Common name
Compositae	*Artemisia* spp.	Wormwood
Compositae	*Balsamorhiza sagittata* (Pursh) Nuttall	Balsamroot
Compositae	*Bidens pilosa* L.	Bur-marigold
Compositae	*Centaurea americana* Nuttall	Basketflower
Compositae	*Cirsium texanum* Buckley	Texas thistle
Compositae	*Cirsium virginianum* L.	Purple-flowered thistle
Compositae	*Dahlia pinnata* Cavanilles	Dahlia
Compositae	*Encelia farinosa* A. Gray	Brittlebush
Compositae	*Erigeron annuus* (L.) Persoon	Daisy fleabane White-top fleabane
Compositae	*Erigeron canadensis* L.	Horseweed, Fleabane
Compositae	*Geraea canescens* Torrey & A. Gray	Desert sunflower
Compositae	*Gutierrezia* spp.	Sheepweed, Snakeweed, Brownweed
Compositae	*Helianthus annuus* L.	Sunflower
Compositae	*Heterotheca subaxillaris* (Lamarck) Britten & Rusby [*Heterotheca psammophila* Wagenknecht]	Camphorweed
Compositae	*Lactuca canadensis* L.	Wild lettuce
Compositae	*Lactuca sativa* L.	Lettuce
Compositae	*Lactuca serriola* L. [*Lactuca scariola* L.]	Prickly lettuce
Compositae	*Machaeranthera tanacetifolia* (Humboldt, Bonpland, Kunth) Nees	Tahoka daisy
Compositae	*Sonchus oleraceous* L.	Sowthistle
Compositae	*Vernonia altissima* Nuttall	Ironweed
Compositae	*Xanthium strumarium* L.	Common cocklebur
Cornaceae	*Cornus drummondi* C.A. Meyer	Rough-leaf dogwood
Cornaceae	*Cornus florida* L.	Flowering dogwood
Cruciferae	*Brassica kaber* (de Candolle) Wheeler	Common mustard
Cruciferae	*Brassica napus* L.	Rape, Rutabaga
Cruciferae	*Brassica nigra* (L.) Koch	Black mustard
Cruciferae	*Brassica oleracea* L.	Cabbage, Kale, Kholrabi, Collard
Cruciferae	*Brassica oleracea* L. var. *acephala* de Candolle	Kale
Cruciferae	*Brassica oleracea* L. var. *botrytis* L.	Broccoli, Cauliflower
Cruciferae	*Brassica oleracea* L. var. *capitata* L.	Cabbage
Cruciferae	*Brassica oleracea* L. var. *gongylodes* L.	Kohlrabi
Cruciferae	*Brassica rapa* L.	Turnip, Field mustard
Cruciferae	*Brassica* spp.	Yellow mustard
Cruciferae	*Capsella bursa-pastoris* (L.) Medicus	Shepherd's purse
Cruciferae	*Cardamine pratensis* L.	Bitter cress

TABLE 2.4 (CONTINUED)
Family, Scientific Name, and Common Name of Each Plant or Crop Mentioned Where Stink Bug Feeding and Development Were Observed or Implied, Alphabetized by Family[a]

Family[b]	Scientific Name	Common name
Cruciferae	*Descurainia pinnata* (Walter) Britton	Tansy mustard
Cruciferae	*Lepidium virginicum* L.	Peppergrass, Peppercress
Cruciferae	*Raphanus raphanistrum* L.	Wild radish
Cruciferae	*Raphanus sativus* L.	Radish
Cruciferae	*Rorippa islandica* (Oeder) Borbas	Bog marsh cress
Cruciferae	*Sisymbrium altissimum* L.	Tumble mustard
Cruciferae	*Sisymbrium irio* L.	London rocket
Cucurbitaceae	*Cucurbita maxima* Duchesne	Squash
Cupressaceae	*Juniperus communis* L.	Mountain juniper, Juniper
Cupressaceae	*Juniperus horizontalis* Moench	Creeping juniper
Cyperaceae	*Carex* spp.	Sedge
Cyperaceae	*Cyperus esculentus* L.	Nutsedge
Cyperaceae	*Rhynchospora inexpansa* (Michaux) Vahl	Nodding beak rush
Ephedraceae	*Ephedra aspera* Engelmann	Mormon tea
Ephedraceae	*Ephedra* spp.	Mormon tea
Ericaceae	*Arctostaphylos pungens* Kunth	Pointleaf manzanita
Geraniaceae	*Erodium cicutarium* (L.) L'Heritier	Alfileria
Gramineae	*Avena barbata* Brotero	Slender wild oats
Gramineae	*Avena fatua* L.	Tartarian wild oats
Gramineae	*Avena sativa* L.	Oats
Gramineae	*Bouteloua* spp.	Gramma
Gramineae	*Brachiaria fasciculata* (Swartz) Parodi	Browntop-millet
Gramineae	*Brachiaria texana* Buckley	Texas millet
Gramineae	*Cynodon dactylon* (L.) Persoon	Bermuda grass
Gramineae	*Digitaria sanguinalis* (L.) Scopoli	Crab grass
Gramineae	*Echinochloa colona* (L.) Link	Junglerice
Gramineae	*Echinochloa crusgalli* (L.) Beauvois	Barnyard grass
Gramineae	*Festuca* spp.	Fescue
Gramineae	*Glyceria septentrionalis* Hitchcock	Eastern manna-grass, Sweet grass
Gramineae	*Hordeum vulgare* L.	Barley
Gramineae	*Lolium multiflorum* Lamarck	Italian ryegrass
Gramineae	*Lolium perenne* L.	Perennial ryegrass
Gramineae	*Oryza sativa* L.	Rice
Gramineae	*Panicum coloratum* L.	Klein grass
Gramineae	*Panicum dichotomiflorum* Michaux	Fall panicum
Gramineae	*Paspalum dilatatum* Poiret	Dallis grass
Gramineae	*Paspalum distichum* L.	Knot grass
Gramineae	*Panicum fasciculatum* Swartz	Browntop panicum
Gramineae	*Panicum hians* Elliott	Gaping panicum
Gramineae	*Paspalum longipilum* Nash	Paspalum
Gramineae	*Paspalum pubiflorum* Ruprecht	Paspalum

TABLE 2.4 (CONTINUED)
Family, Scientific Name, and Common Name of Each Plant or Crop Mentioned Where Stink Bug Feeding and Development Were Observed or Implied, Alphabetized by Family[a]

Family[b]	Scientific Name	Common name
Gramineae	*Paspalum urvillei* Steudel	Vasey grass
Gramineae	*Phalaris minor* Retzius	Little-seed canary grass
Gramineae	*Poa annua* L.	Annual blue grass
Gramineae	*Saccharum officinarum* L.	Sugarcane
Gramineae	*Secale cereale* L.	Rye
Gramineae	*Setaria italica* (L.) Beauvois	Foxtail millet, Foxtail grass
Gramineae	*Sorghum bicolor* (L.) Moench	Grain sorghum, Milo, Feterita, Sorghum, Milo maize
Gramineae	*Sorghum halepense* (L.) Persoon	Johnson grass
Gramineae	*Sorghum sudanese* (Piper) Stapf	Sudan grass
Gramineae	*Sorghum vulgare* Persoon	Kafir corn, Broom corn
Gramineae	*Stipa avenacea* L.	Oatgrass
Gramineae	*Triticum aestivum* L.	Wheat, Spelt
Gramineae	*Triticum turgidum* L.	English wheat, Winter emmer
Gramineae	*Triticum vulgare* Villars	Wheat
Gramineae	*Zea mays* L.	Corn (field, sweet)
Grossulariaceae	*Ribes cereum* Douglas	Currant, Wax
Juglandaceae	*Carya illinoensis* (Wangenheim) K. Koch	Pecan
Labiatae	*Marrubium vulgare* L.	Horehound
Labiatae	*Salvia officinalis* L.	Sage
Leguminosae	*Albizia julibrissin* Durazzini	Mimosa
Leguminosae	*Arachis hypogaea* L.	Peanut
Leguminosae	*Cassia leptocarpa* Bentham	Slimpod senna
Leguminosae	*Cercis* spp.	Redbud
Leguminosae	*Chamaecrista fasciculata* (Michaux) Greene [*Cassia fasciculata* Michaux]	Partridge pea
Leguminosae	*Crotalaria brevidens* Bentham	Slender-leaf crotalaria
Leguminosae	*Crotalaria lanceolata* E. Meyer	Lance-leaf crotalaria
Leguminosae	*Crotalaria spectabilis* Roth	Showy crotalaria
Leguminosae	*Desmodium tortuosum* (Swartz) de Candolle	Beggerweed
Leguminosae	*Gleditsia triacanthos* L.	Honey locust
Leguminosae	*Glycine max* (L.) Merrill	Soybean
Leguminosae	*Indigofera hirsuta* L.	Hairy indigo
Leguminosae	*Lens culinaris* Medicus	Lentil
Leguminosae	*Medicago hispida* Gaertn	Bur-clover
Leguminosae	*Medicago sativa* L.	Alfalfa
Leguminosae	*Melilotus indica* (L.) Allioni	Indian sweet-clover, Alfalfilla, Sour-clover
Leguminosae	*Phaseolus lunatus* L.	Lima bean
Leguminosae	*Phaseolus vulgaris* L.	Snap bean, Bean, Green bean
Leguminosae	*Pisum sativum* L.	Pea (field, English)
Leguminosae	*Prosopis juliflora* (Swartz) de Candolle	Mesquite

TABLE 2.4 (CONTINUED)
Family, Scientific Name, and Common Name of Each Plant or Crop Mentioned Where Stink Bug Feeding and Development Were Observed or Implied, Alphabetized by Family[a]

Family[b]	Scientific Name	Common name
Leguminosae	*Prosopis velutina* Wooton	Velvet mesquite
Leguminosae	*Robinia pseudo-acacia* L.	Black locust
Leguminosae	*Senna occidentalis* (L.) Link [*Cassia occidentalis* L.]	Coffee senna
Leguminosae	*Sesbania emerus* (Aublet) Urban	Prairie bean
Leguminosae	*Stizolobium deeringianum* Bort	Velvet bean
Leguminosae	*Trifolium* spp.	Clover
Leguminosae	*Vicia angustifolia* Reichard	Vetch
Leguminosae	*Vigna radiata* (L.) Wilczek	Mung bean
Leguminosae	*Vigna unguiculata* (L.) Walpers [*Vigna sinensis* (L.) Endlicher]	Cowpea, Southern pea
Liliaceae	*Allium cepa* L.	Onion
Liliaceae	*Asparagus officinalis* L.	Asparagus
Linaceae	*Linum usitatissimum* L.	Flax, Linseed
Malvaceae	*Abelmoschus esculentus* (L.) Moench	Okra
Malvaceae	*Gossypium hirsutum* L.	Cotton
Malvaceae	*Malva parviflora* L.	Mallow, Cheeseweed
Malvaceae	*Sphaeralcea emoryi* Torrey ex A. Gray	Globe-mallow
Moraceae	*Ficus carica* L.	Fig
Oleaceae	*Ligustrum sinense* Loureiro	Chinese privet
Onagraceae	*Gaura parviflora* Douglas ex Lehmann	Gaura, lizard's tail, velvetweed
Onagraceae	*Gaura villosa* Torrey	Woolly gaura, Hairy gaura
Passifloraceae	*Passiflora incarnata* L.	May-pop
Pedaliaceae	*Sesamum indicum* L.	Sesame
Pinaceae	*Picea* spp.	Spruce
Pinaceae	*Pinus contorta* Douglas	Lodgepole pine
Pinaceae	*Pinus ponderosa* Douglas	Ponderosa pine
Pinaceae	*Pinus sylvestris* L.	Scots' pine (Scotch pine)
Plantaginaceae	*Plantago major* L.	Broad-leaved plantain
Polygonaceae	*Eriogonum* spp.	Wild buckwheat
Polygonaceae	*Fagopyrum esculentum* Moench	Buckwheat
Polygonaceae	*Rumex crispus* L.	Curlyleaf dock, Curly dock
Polygonaceae	*Rumex hymenosepalus* Torrey	Canaigre
Polygonaceae	*Rumex occidentalis* S. Watson	Western dock
Proteaceae	*Macadamia integrifolia* Maiden & Betche	Macadamia
Rosaceae	*Dryas drummondii* Richardson	Dryas
Rosaceae	*Fragaria vesca* L.	European strawberry
Rosaceae	*Fragaria* spp.	Strawberry
Rosaceae	*Malus domestica* Borkhauser	Apple
Rosaceae	*Photinia serrulata* L.	Photinia
Rosaceae	*Prunus armeniaca* L.	Apricot
Rosaceae	*Prunus avium* L.	Cherry (sweet)

TABLE 2.4 (CONTINUED)
Family, Scientific Name, and Common Name of Each Plant or Crop Mentioned Where Stink Bug Feeding and Development Were Observed or Implied, Alphabetized by Family[a]

Family[b]	Scientific Name	Common name
Rosaceae	*Prunus domestica* L.	Plum
Rosaceae	*Prunus persica* (L.) Batsch	Peach
Rosaceae	*Prunus serotina* Ehrhart	Wild black cherry, Black cherry
Rosaceae	*Purshia tridentata* (Pursh) de Candolle	Bitterbrush
Rosaceae	*Pyrus communis* L.	Pear
Rosaceae	*Rubus flagellaris* Willdenow	American dewberry
Rosaceae	*Rubus idaeus* L.	Red raspberry, Raspberry
Rosaceae	*Rubus laciniatus* Willdenow	Cut-leaf blackberry, Blackberry
Rutaceae	*Citrus sinensis* (L.) Osbeck	Orange
Rutaceae	*Citrus reticulata* Blanco	Tangerine
Scrophulariaceae	*Verbascum thapsus* L.	Woolly mullein, Common mullein
Solanaceae	*Capsicum annuum* L.	Bell pepper, Pepper
Solanaceae	*Datura stramonium* L.	Jimsonweed
Solanaceae	*Lycium pallidum* Miers	Desert thorn
Solanaceae	*Lycopersicon esculentum* Miller	Tomato
Solanaceae	*Nicotiana tabacum* L.	Tobacco
Solanaceae	*Solanum elaeagnifolium* Cavanilles	White horse-nettle
Solanaceae	*Solanum melongena* L.	Eggplant
Solanaceae	*Solanum tuberosum* L.	Potato
Tiliaceae	*Tilia* spp.	Linden
Vitaceae	*Ampelopsis arborea* (L.) Koehne	Peppervine
Vitaceae	*Vitis labrusca* L.	Fox grape (concord)
Vitaceae	*Vitis* spp.	Wild grape, Grape
Zygophyllaceae	*Larrea tridentata* (Sesse & Mocino ex de Candolle) Coulter	Creosote-bush

[a] All scientific names were verified with one or more authoritative references including Fernald 1950, Bailey and Bailey 1976, Radford et al. 1981, Faust and Strang 1983, Weed Sci. Soc. Am. 1984, and Brako et al. 1997.

[b] Depending on taxonomic source, some plant family names are different than those appearing in this table. These names include Lamiaceae (= Labiatae), Fabaceae (= Leguminosae), Poaceae (= Gramineae), Asteraceae (= Compositae), Brassicaceae (= Cruciferae), and Simmondiaceae (= Buxaceae) (Brako et al. 1997). Although five of these names do not end in -aceae, they have been sanctioned by the International Code of Botanical Nomenclature because of old, traditional usage (Jones and Luchsinger 1986).

3 Superfamily Pentatomoidea Leach: Keys for Identification and Introduction to Pest Species

CONTENTS

3.1   INTRODUCTION

The following keys include those pentatomine stink bugs that we consider to be of primary economic importance and other pentatomoid taxa (including other stink bugs) likely to be collected in the field concurrently. Those of economic importance are discussed in detail in Chapters 4 through 11.

For each stink bug pest, information is included on North American geographic distribution, general economic importance, life history, laboratory investigations, and references to descriptions of the immature stages. Following that is a discussion of damage caused by each stink bug to selected crops including, where possible, the relationship between the bug's life history and the resulting damage, and a discussion of the damage itself. The host plants, both wild and cultivated, for each stink bug are listed in Tables 2.2 to 2.4.

3.2   KEY TO FAMILIES (MODIFIED FROM McPHERSON 1982)

    1.   Scutellum large, U-shaped, longer than corium, almost covering entire abdomen (Figure 3.3) 2

1'. Scutellum smaller, subtriangular, usually shorter than corium (Figure 3.1); if scutellum large and U-shaped, then colors bright and contrasting or prominent tooth or process present at each anterolateral angle of pronotum (Figure 3.5) 3

2(1). Tibiae each with 2 or more rows of stout spines (Figure 3.4); color basically black, hemelytra each may have yellow marking along costal margin of corium (negro bugs) .... Thyreocoridae Amyot and Serville

2'. Tibiae without stout spines (Figure 3.6); color variable but never shiny black (shieldbacked bugs) .... Scutelleridae Leach

3(1'). Tibiae with stout spines (Figure 3.7); front legs often strongly fossorial (Figure 3.7); length rarely more than 7 mm (burrower bugs) .... Cydnidae Billberg

3'. Tibiae without stout spines; front legs not fossorial; length usually more than 7 mm 4

4(3'). Tarsi 2-segmented; thoracic and abdominal sterna with conspicuous, mid-ventral longitudinal ridge (acanthosomatids) .... Acanthosomatidae Signoret

4'. Tarsi 3-segmented; thoracic and abdominal sterna without conspicuous, longitudinal ridge (stink bugs) .... Pentatomidae Leach

3.3 FAMILY PENTATOMIDAE LEACH

The following keys include stink bugs that are noted pest species, those that are not pests but often are encountered, and those that rarely are encountered or are predators; those in the latter two categories will key only to the subfamily. More detailed keys for these latter taxa are available elsewhere (see Blatchley 1926, McPherson 1982, Slater and Baranowski 1978, and references therein).

3.3.1 KEY TO SUBFAMILIES (MODIFIED FROM MCPHERSON 1982)

1. Eyes prominent, pedunculate (Figure 3.5); scutellum U-shaped, reaching almost to tip of abdomen; frena less than one fourth as long as scutellum (turtle bugs) (Figure 3.5) .... Podopinae Amyot and Serville

1'. Eyes not pedunculate (Figures 3.1 and 3.2); scutellum usually subtriangular but if large and U-shaped, then colors bright and contrasting; frena at least one fourth as long as scutellum (Figure 3.1) 2

2(1'). Metasternum produced anteriorly as bifid structure (Figure 3.8); beak short, not surpassing middle coxae (Figure 3.8)....... Edessinae Kirkaldy

2'. Metasternum not as above; beak usually reaching or surpassing hind coxae (Figure 3.2) 3

3(2'). Beak with segment 1 short, thick, free, with, at most, only base between bucculae, which are united posteriorly (Figure 3.9) (predaceous stink bugs) .... Asopinae Spinola

3'. Beak with segment 1 slender, lying between bucculae, which are subpar-
·   allel and not united posteriorly (Figures 3.2, 3.14, 3.18, 3.22, 3.47)
     ....Pentatominae Leach

3.4 SUBFAMILY PENTATOMINAE LEACH[1,2]

3.4.1 KEY TO SPECIES AND SUBSPECIES

1.   Juga each with subapical tooth on outer side (Figures 3.10 and 3.11);
     pronotum with anterolateral margins coarsely dentate (Figures 3.10 and
     3.11); abdominal venter with shallow, longitudinal, median groove that
     becomes obsolete posteriorly 2

1'. Juga without subapical teeth (Figures 3.1 and 3.2); pronotum with ante-
     rolateral margins variable but usually not coarsely dentate (Figures 3.1
     and 3.2); abdominal venter usually without median groove although
     occasionally with slight indentation anteriorly......................................4

2(1). Scutellum with basal third to fourth elevated above remainder; humeri
       quadrate with large teeth (Figure 3.10)
       ....*Parabrochymena arborea* (Say)

2'. Scutellum with basal third to fourth only slightly elevated above remain-
     der; humeri subtriangular with small teeth (Figure 3.11)
     (*Brochymena*)..3

3(2'). Juga subfoliate, converging to overlapping apically, leaving no preapical
        sinus in front of tylus (Figure 3.11); humeri obtuse (Figure 3.11)
        .... *B. quadripustulata* (Fab.)

3'. Juga not subfoliate, their inner margins diverging to slightly converging,
     usually not touching, leaving preapical sinus in front of tylus; humeri
     acute..*B. sulcata* Van Duzee

4(1'). Abdominal venter with stridulatory areas on first 3 sternites each side of
        middle; body elongate, length about 4 times greatest width (Figure 3.12)
        ....*Mecidea major* Sailer

4'. Abdominal venter without stridulatory areas on first 3 sternites each side
     of middle; length not more than 3 times greatest width 5

5(4'). Abdominal sternite 2 armed medially with anteriorly directed spine or
        tubercle (except in male *Dendrocoris humeralis*) (Figures 3.2, 3.14, 3.17,
        3.18, 3.22)..6

5'. Abdominal sternite 2 unarmed...13

6(5). Juga longer than tylus, may be contiguous in front of it (Figure 3.13)....7

6'. Juga usually not longer than tylus but, if so, not contiguous in front of
     it (Figure 3.15)...8

7(6). Juga with tips acute, not contiguous in front of tylus; humeri spinose;
       mesosternum with prominent, acute, median ridge
       .... *Arvelius albopunctatus* (De Geer)

7'.   Juga with tips obtuse, usually contiguous in front of tylus (Figure 3.13); humeri subacute (Figure 3.13); mesosternum without acute prominent ridge ....*Dendrocoris humeralis* (Uhler)

8(6').   Spine of abdominal sternite 2 reaching or slightly surpassing middle coxae (Figure 3.14); spiracular peritremes black .... *Piezodorus guildinii* (Westwood)

8'.   Spine of abdominal sternite 2 not reaching middle coxae (Figures 3.2, 3.17, 3.18, 3.22); spiracular peritremes pale..9

9(8').   Ostiolar canal short, not extending laterally to middle of supporting plate (Figure 3.16); spine of abdominal sternite 2 short, obtuse (Figure 3.17) .... *Nezara viridula* (L.)

9'.   Ostiolar canal long, extending laterally to beyond middle of its supporting plate (Figures 3.2, 3.18, 3.22, 3.26); spine of abdominal sternite 2 variable 10

10(9').   Smaller, length 12 mm or less; spine of abdominal sternite 2 short, obtuse (Figures 3.18 and 3.22); color variable (*Banasa*)................................11

10'.   Larger, length 12.5 mm or more; spine of abdominal sternite 2 long, acutely tapering (Figure 3.2); color green (*Acrosternum*).....................12

11(10).   Abdominal sternites often with prominent black spots at lateral angles; male without prominent acute tubercles projecting dorsally from posterior margin of pygophore (Figure 3.19), claspers sublinear in shape (Figure 3.20); female with posterior margin of each basal plate of genital segment noticeably concave (Figure 3.21)........................*B. calva* (Say)

11'.   Abdominal sternites without black spots, or with only minute spots, at lateral angles; male with prominent acute tubercles projecting dorsally from posterior margin of pygophore (Figure 3.23), claspers subtriangular in shape (Figure 3.24); female with posterior margin of each basal plate of genital segment straight to slightly concave (Figure 3.25) .... *B. dimidiata* (Say)

12(10').   Male with posterior margin of pygophore subtruncate, sinuate, with median notch, its outer angles subacute (Figure 3.27); female with posterior margin of each basal plate of genital segment evenly convex (Figure 3.28) ....*A. hilare* (Say)

12'.   Male with posterior margin of pygophore broadly V-shaped, its outer angles obtuse (Figure 3.29); female with posterior margin of each basal plate of genital segment with posterolateral projection, forming obtuse angle (Figure 3.30) .... *A. marginatum* (Palisot de Beauvois)

13(5').   Ostiole usually with distinct auricle (i.e., a small earlobe- or flap-like, free-edged process; not evident in *Murgantia*) and, at most, a short abruptly ending canal (Figure 3.40), ostiole often rounded on inner side 14

13'.   Ostiole without auricle but extended as elongated tapering canal (as in Figure 3.26); ostiole V-shaped on inner side 34

14(13). Ostiole small, without evident auricle or canal; head strongly declivent with front of head almost vertical, sides of juga thickened and reflexed; color variegated black and red or black and yellow
....*Murgantia histrionica* (Hahn)

14'. Ostiole conspicuous, auricle usually well-developed; head varying from horizontal to strongly declivent, sides of juga not thickened and reflexed; color variable but not variegated black and red or black and yellow.... 15

15(14'). Larger, length 17 mm or more; juga longer than tylus, tips acute; humeri spinose..*Loxa flavicollis* (Drury)

15'. Smaller, length 15 mm or less; juga variable in length to tylus; humeri variable.. 16

16(15'). Scutellum shorter than corium, with apical third narrower than apex of corium and with tip not broadly rounded (Figure 3.48)...................... 17

16'. Scutellum subequal to or longer than corium, with apical third wider than apex of corium and with tip broadly rounded (Figures 3.49 and 3.50)...32

17(16). Hind tibiae distinctly sulcate dorsally throughout their lengths (Figure 3.35); tylus rounded at apex, not reflexed................................ 18

17'. Hind tibiae usually not distinctly sulcate throughout their lengths but if shallow longitudinal groove present, then tylus is acute at apex and reflexed (Figures 3.43 and 3.44).. 29

18(17). Pronotum with anterolateral margins crenulate, humeri prominent, usually acute (rounded in *E. t. luridus*) (Figures 3.34, 3.37, 3.38, 3.39) (*Euschistus*)
.... 19

18'. Pronotum with anterolateral margins not crenulate, humeri broadly rounded (Figure 3.31).................................... *Hymenarcys nervosa* (Say)

19(18). Hemelytra with each membrane marked with brownish spots, easily seen against white paper inserted beneath membrane 20

19'. Hemelytra with each membrane not marked with brownish spots 27

20(19). Pronotum with distinct yellowish callous line between humeri........... 21

20'. Pronotum without distinct yellowish callous line between humeri...... 22

21(20). Spiracular peritremes pale; pro- and mesopleura with evaporative areas marked with small reddish brown to black punctures ... *E. crassus* Dallas

21'. Spiracular peritremes black; pro- and mesopleura with evaporative areas not marked with reddish brown to black punctures ....*E. ictericus* (L.)

22(20'). Abdominal sternites with black spots at lateral angles (occasional specimens have light diffuse spots); humeri variable; body form variable; antennae with segments 4 and 5 pale red, or brown, to black; male without black spot on ventral side of pygophore.. 23

22'. Abdominal sternites without black spots at lateral angles (occasional specimens have light diffuse spots); humeri generally acute to spinose; antennae with apical half of segment 4 and all of 5 black; male with dark brown to black spot on ventral side of pygophore
....*E. variolarius* (Palisot de Beauvois)

23(22).   Abdominal venter with 1 to 4 median black spots, which occasionally are obsolete; length less than 12 mm (*E. tristigmus*) 24

23'.   Abdominal venter without median black spots; length variable 25

24(23).   Humeri rounded (Figure 3.37); antennae with apical half of segment 4 and all of 5 brownish black to black (northern U.S.)
....*E. tristigmus luridus* Dallas[3]

24'.   Humeri subtriangular to spinose (Figures 3.38 and 3.39); antennae entirely pale or reddish brown (southern U.S.)
....*E. tristigmus tristigmus* (Say)[3]

25(23').   Legs heavily marked with large brown spots (Figure 3.33); juga subequal in length to tylus (Figure 3.32) ....*E. conspersus* Uhler

25'.   Legs marked with small brown to reddish brown spots (occasional specimens may have large spots) (Figure 3.36); juga variable in length relative to tylus (*E. servus*)...26

26(25').   Juga equal or subequal in length to tylus; antennae entirely yellowish brown or reddish brown; connexiva broadly exposed (southern U.S.)
....*E. servus servus* (Say)[4]

26'.   Juga distinctly longer than tylus; antennae with segments 4 and 5 usually dark brown; connexiva nearly or completely covered by hemelytra (northern U.S.)....................................*E. servus euschistoides* (Vollenhoven)[4]

27(19').   Hemelytra each with numerous black spots on corium; pronotum with conspicuous calloused line between humeri; head and thoracic area anterior to pronotal stripe heavily punctate, darker than remainder of dorsum
....*E. obscurus* (Palisot de Beauvois)

27'.   Hemelytra each usually without black spots on corium; pronotum without conspicuous calloused line between humeri; head and pronotum not punctate and colored as above 28

28(27').   Spiracular peritremes pale; hemelytra each with membrane white to light brown against white paper inserted beneath membrane
....*E. crenator* (Fab.)

28'.   Spiracular peritremes black; hemelytra each with membrane distinctly brown ....*E. biformis* Stål

29(17').   Humeri rounded (Figure 3.42); beak with segment 1 distinctly longer than bucculae (Figure 3.41); pronotum and scutellum, except at base, with margins yellow, calloused (Figure 3.42); pronotum with yellow calloused transverse line on anterior half (Figure 3.42)
....*Mormidea lugens* (Fab.)

29'.   Humeri acute (Figures 3.43, 3.46, 3.48); beak with segment 1 not longer than bucculae (Figure 3.45); pronotum and scutellum not marked as above 30

30(29').   Color generally yellow to yellowish brown; tylus rounded at apex, not reflexed, equaling or slightly exceeding juga (*Oebalus*) (Figures 3.46 and 3.48) 31

30'. Color generally dark brown to black; tylus acute at apex, reflexed, strongly surpassing juga (Figures 3.43 and 3.44) .... *Proxys punctulatus* (Palisot de Beauvois)

31(30). Humeri spinose, spines strongly directed anteriorly (Figure 3.46); antennae with segment 2 longer than 1; scutellum without wide calloused mark each side basally on lateral margin (Figure 3.46); spiracular peritremes black .... *O. pugnax pugnax* (Fab.)

31'. Humeri acute to spinose, spines directed laterally (Figure 3.48); antennae with segment 2 shorter than 1 (Figure 3.47); scutellum with wide calloused mark each side basally on lateral margin (Figure 3.48); spiracular peritremes pale .... *O. ypsilongriseus* (De Geer)

32 (16'). Juga longer than tylus and contiguous in front of it or nearly so (Figure 3.49) .... *Aelia americana* Dallas

32'. Juga subequal to tylus (Figure 3.50) 33

33(32). Body dull yellow; hemelytra each with reticulate venation in membrane .... *Coenus delius* (Say)

33'. Body black with red or reddish yellow markings; hemelytra each without reticulate venation in membrane .... *Cosmopepla lintneriana* Kirkaldy[5]

34(13'). Juga longer than tylus by distance equal to at least width of tylus at apex, tips rounded, converging and frequently contiguous (*Holcostethus*) (Figure 3.51) 35

34'. Juga slightly shorter to slightly longer than tylus but if longer, then exceeding tylus by less than width of tylus at apex (Figures 3.52, 3.53, 3.55, 3.57, 3.59, 3.61) 36

35(34). Connexivum black, bordered by narrow yellow margin; antennal segments 1 to 3 yellowish to reddish, 4 and 5 reddish brown to brownish black .... *H. limbolarius* (Stål)

35'. Connexivum alternated with black and yellow, black spots at lateral angles reaching or almost reaching edge; antennal segments often more uniformly colored .... *H. abbreviatus* Uhler

36(34'). Body distinctly pubescent, especially sides of abdominal sternites; frena reaching about middle of scutellum .... *Trichopepla semivittata* (Say)

36'. Body not pubescent or only very slightly so; frena surpassing middle of scutellum 37

37(36'). Ostiolar canal extending laterally beyond middle of supporting plate almost to outer front angle (*Thyanta*) 38

37'. Ostiolar canal not extending laterally beyond middle of supporting plate (*Chlorochroa*) 42

38(37). Humeral angles spinose, spines directed anterolaterad to almost 45° with longitudinal axis of body (Figure 3.52) .... *T. perditor* (Fab.)

38'. Humeral angles variable but if spinose, spines not directed anterolaterad to almost 45° with longitudinal axis of body (Figures 3.53 and 3.55) ...39

39(38').   Anterolateral margins of pronotum, lateral angles of abdominal sternites, ventral rows of postspiracular spots, and inner angle of each pronotal cicatrice, black 40

   39'.   Anterolateral margins of pronotum not black; other black markings variable, often absent..41

40(39).   Pronotum with black borders of anterolateral margins covering both dorsal and ventral sides (Figure 3.53); pygophore with median notched lobe subtriangular and subtended by vague semicircular impression on the cup (Figure 3.54) .... *T. calceata* (Say)

   40'.   Pronotum with black borders of anterolateral margins limited to ventral side (Figure 3.55); pygophore with median notched lobe nearly straight-edged and subtended by clearly defined, rectangular impression on the cup (Figure 3.56) ....*T. custator custator* (Fab.)

41(39').   Humeri rounded to spinose; abdominal sternites each with black postspiracular spot each side, which often is larger in diameter than adjacent spiracle, or spot absent; posterolateral angle of each abdominal sternite with or without minute black spot..............*T. custator accerra* McAtee[6]

   41'.   Humeri rounded to subacute; abdominal sternites with postspiracular spots usually absent but if present, then usually smaller in diameter than adjacent spiracles; posterolateral angle of each abdominal sternite without black spot (western species)...................................*T. pallidovirens* (Stål)[6]

42(37').   Embolium subparallel throughout much of its length to slightly widened distally (Figures 3.57, 3.58, 3.59, 3.60); scutellum with three large to small distinct pale callosities at base (Figures 3.57 and 3.59); paramere with angle between median and posterior processes more than 90° ... 43

   42'.   Embolium noticeably widened distally (Figures 3.61 and 3.62); scutellum with pale callosites either very small or absent (Figure 3.61); paramere with angle between median and posterior processes about 90°...........44

43(42).   Scutellum with callosities at base large, distinct (Figure 3.57); hemelytra each with membrane marked with purplish flecks .... *C. sayi* (Stål)[7]

   43'.   Scutellum with callosities at base moderate to small (Figure 3.59); hemelytra each with membrane not marked with purplish flecks .... *C. uhleri* (Stål)[7]

44(42').   Color usually light green; pronotum with anterolateral margins white to yellowish pink; scutellum usually white to yellow at apex; pygophore with well-developed transverse subapical ridge ventrally (Figure 3.64) ....*C. granulosa* (Uhler)[7]

   44'.   Color olive-green to black; pronotum with anterolateral margins pale orange to red; scutellum pale orange to red at apex; pygophore with ventral, transverse subapical ridge interrupted by distinct median tumescence (Figure 3.63) .... *C. ligata* (Say)[7]

---

[1] The first abdominal sternite is hidden and, therefore, the first visible sternite actually is the second. To simplify use of this key, numbering of abdominal sternites refers to sternites that are visible (e.g., the anteriorly directed spine found in some stink bugs on the third morphological sternite is referred to in our key as arising from the second).

[2] Although this subfamily contains four tribes (i.e., Sciocorini Amyot and Serville, Halyini Dallas, Mecideini Distant, and Pentatomini Leach), all species of major economic importance considered in this work are within the Pentatomini; others in the tribe are not economically important but often collected in small numbers. Species within the Sciocorini are rare, and those within the Mecideini are not of economic importance. Several species of Halyini are widely distributed, particularly *Brochymena quadripustulata* and *Parabrochymena arborea* in the eastern three fourths of North America and *Brochymena sulcata* in the western United States. To increase the usefulness of the key, we have included those species that often are collected (including *Mecidea major* and the three halyines), whether or not they are of economic importance.

[3] The taxonomic status and distribution of this subspecies are discussed by McPherson (1982).

[4] The taxonomic status and distribution of this subspecies are discussed by Sailer (1954) and McPherson (1982). *E. servus euschistoides* and *E. s. servus* meet in a broad zone of intergradation from Maryland to Kansas where specimens occur that have combinations of characters of both subspecies.

[5] This species usually has been referred to as *Cosmopepla bimaculata* (Thomas), a name that now is considered invalid (see discussion by Rider and Rolston [1995]).

[6] *T. custator accerra* and *T. pallidovirens* are genetically distinct, based on chromosome number. Unfortunately, no morphological character will distinguish them consistently. *T. custator accerra* occurs over most of the United States, being absent primarily in Florida and much of the Northwest. *T. pallidovirens* occurs in the western United States and is common in the Northwest. Where the taxa overlap in the southwestern United States, *T. custator accerra* usually has angulate to spinose humeri, whereas *T. pallidovirens* has rounded humeri. *T. custator accerra* apparently does not occur in Washington, Oregon, or Idaho, whereas *T. pallidovirens* is common. (See Rider and Chapin [1992] for a more detailed discussion of the taxonomy of these taxa.)

[7] The taxonomy of this species, and others in *Chlorochroa*, is discussed in more detail by Buxton et al. (1983).

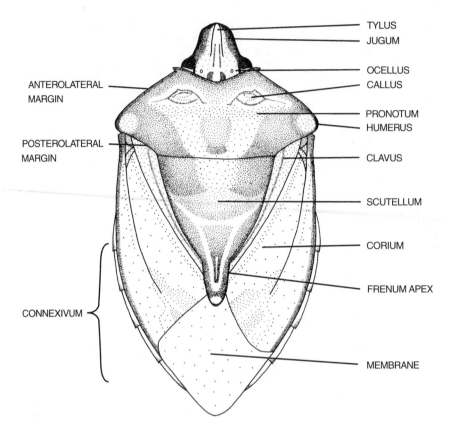

FIGURE 3.1 General dorsal view, *Acrosternum hilare*. The frenum is a groove along each lateral margin of the scutellum, the distal end of which is labeled here as frenum apex.

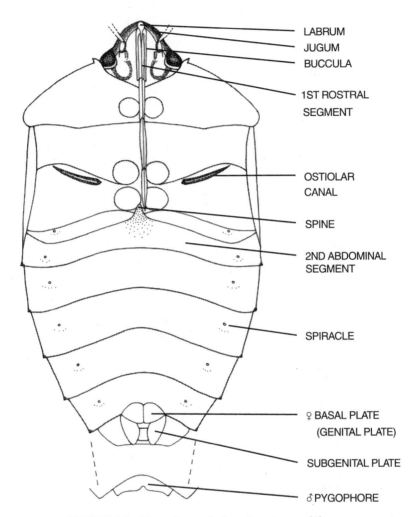

FIGURE 3.2 General ventral view, *Acrosternum hilare*.

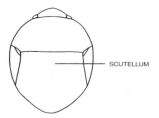

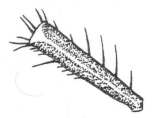

FIGURE 3.3 General dorsal view, *Galgupha carinata.* (Redrawn from McPherson, J.E., The Pentatomoidea (Hemiptera) of northeastern North America with emphasis on the fauna of Illinois, Southern Illinois University Press, Carbondale and Edwardsville, 1982. With permission.)

FIGURE 3.4 Front tibia, *Galgupha denudata.*

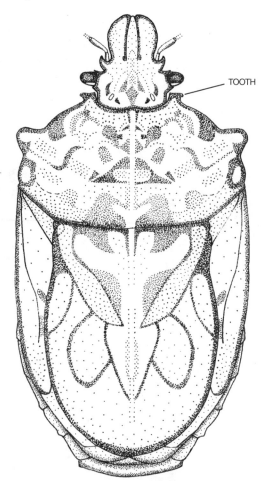

FIGURE 3.5 General dorsal view, *Amaurochrous cinctipes.*

FIGURE 3.6   Front tibia, *Chelysomidea guttata.*

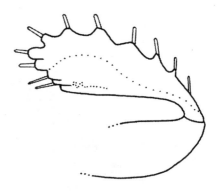

FIGURE 3.7   Front tibia, *Cyrtomenus ciliatus.*

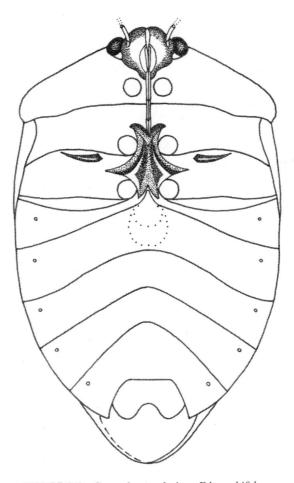

FIGURE 3.8   General ventral view, *Edessa bifida.*

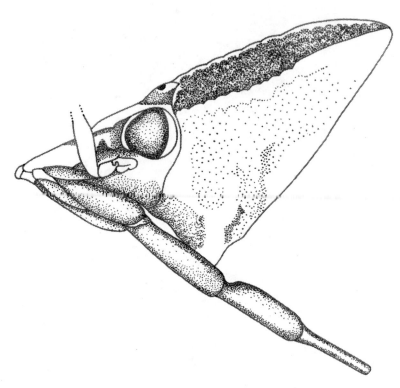

FIGURE 3.9 Head and prothorax, lateral view, *Podisus maculiventris.*

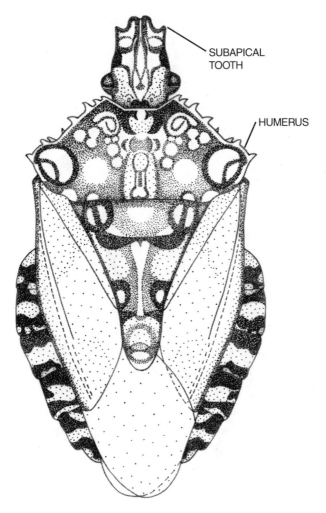

FIGURE 3.10   General dorsal view, *Parabrochymena arborea*.

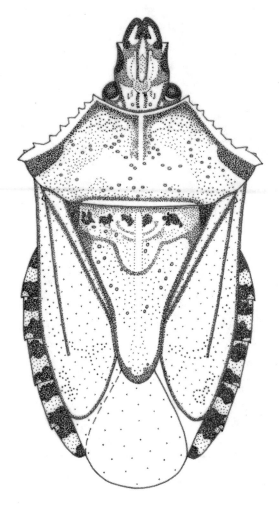

FIGURE 3.11 General dorsal view, *Brochymena quadripustulata.*

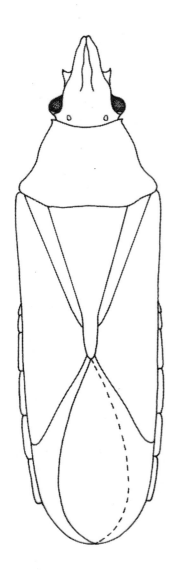

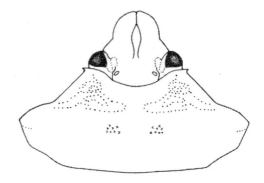

FIGURE 3.13 Head and prothorax, dorsal view, *Dendrocoris humeralis.*

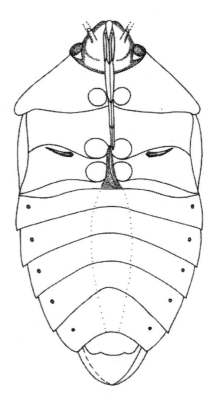

FIGURE 3.12 General dorsal view, *Mecidea major.* (Redrawn from McPherson, J.E., The Pentatomoidea (Hemiptera) of northeastern North America with emphasis on the fauna of Illinois, Southern Illinois University Press, Carbondale and Edwardsville, 1982. With permission.)

FIGURE 3.14 Male, general ventral view, *Piezodorus guildinii.*

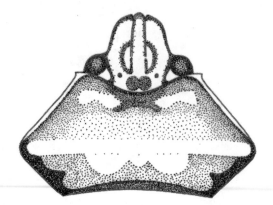

FIGURE 3.15   Head and prothorax, dorsal view, *Piezodorus guildinii.*

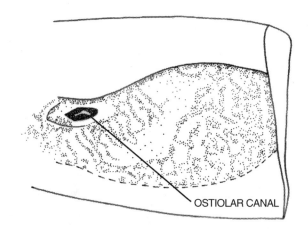

FIGURE 3.16   Metapleuron, *Nezara viridula.*

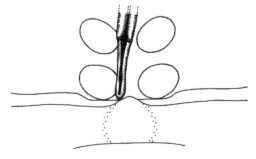

FIGURE 3.17   Spine of second abdominal sternite, *Nezara viridula.*

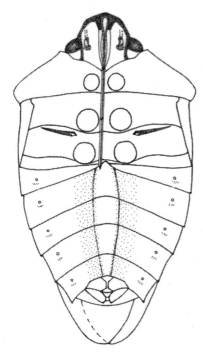

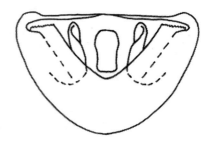

FIGURE 3.19 Pygophore, posterodorsal view, *Banasa calva*. (Redrawn from McPherson, J.E., The Pentatomoidea (Hemiptera) of northeastern North America with emphasis on the fauna of Illinois, Southern Illinois University Press, Carbondale and Edwardsville, 1982. With permission.)

FIGURE 3.18 Female, general ventral view, *Banasa calva*.

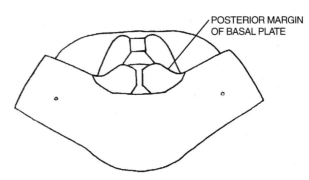

FIGURE 3.20
Left clasper, lateral view, *Banasa calva*. (Redrawn from McPherson, J.E., The Pentatomoidea (Hemiptera) of northeastern North America with emphasis on the fauna of Illinois, Southern Illinois University Press, Carbondale and Edwardsville, 1982. With permission.)

FIGURE 3.21 Female external genitalia and abdominal sternite 6, ventral view, *Banasa calva*. (Redrawn from McPherson, J.E., The Pentatomoidea (Hemiptera) of northeastern North America with emphasis on the fauna of Illinois, Southern Illinois University Press, Carbondale and Edwardsville, 1982. With permission.)

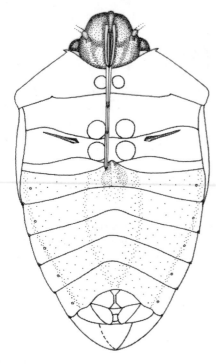

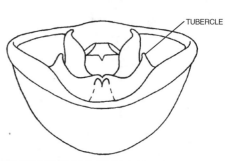

FIGURE 3.23 Pygophore, posterodorsal view, *Banasa dimidiata*. (Redrawn from McPherson, J.E., The Pentatomoidea (Hemiptera) of northeastern North America with emphasis on the fauna of Illinois, Southern Illinois University Press, Carbondale and Edwardsville, 1982. With permission.)

FIGURE 3.22 Female, general ventral view, *Banasa dimidiata*.

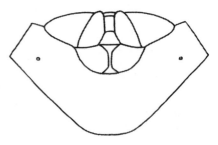

FIGURE 3.24
Left clasper, lateral view, *Banasa dimidata*. (Redrawn from McPherson, J.E., The Pentatomoidea (Hemiptera) of northeastern North America with emphasis on the fauna of Illinois, Southern Illinois University Press, Carbondale and Edwardsville, 1982. With permission.)

FIGURE 3.25 Female external genitalia and abdominal sternite 6, ventral view, *Banasa dimidiata*. (Redrawn from McPherson, J.E., The Pentatomoidea (Hemiptera) of northeastern North America with emphasis on the fauna of Illinois, Southern Illinois University Press, Carbondale and Edwardsville, 1982. With permission.)

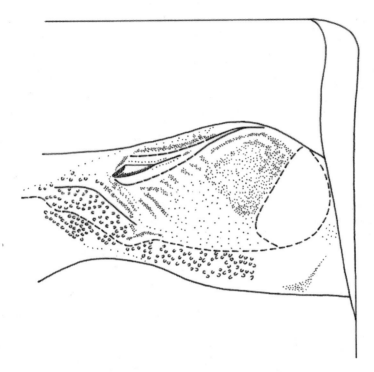

FIGURE 3.26 Metapleuron, *Acrosternum hilare.*

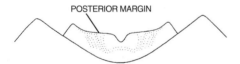

FIGURE 3.27 Pygophore, ventral view, *Acrosternum hilare.*

FIGURE 3.29 Pygophore, ventral view, *Acrosternum marginatum.*

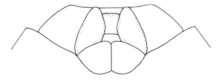

FIGURE 3.28 Female external genitalia, ventral view, *Acrosternum hilare.*

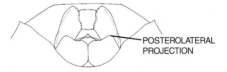

FIGURE 3.30 Female external genitalia, ventral view, *Acrosternum marginatum.*

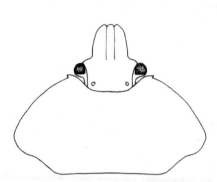

FIGURE 3.31 Head and prothorax, dorsal view, *Hymenarcys nervosa*.

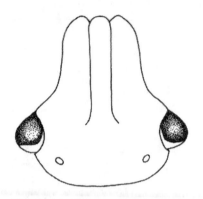

FIGURE 3.32 Head, dorsal view, *Euschistus conspersus*.

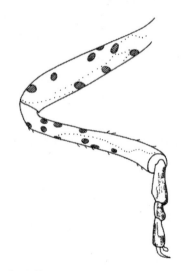

FIGURE 3.33 Front leg, *Euschistus conspersus*.

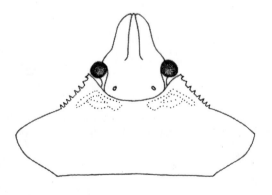

FIGURE 3.34 Head and prothorax, dorsal view, *Euschistus servus*.

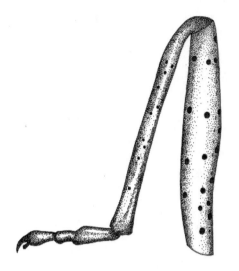

FIGURE 3.35 Hind leg, *Euschistus servus*.

FIGURE 3.36 Front leg, *Euschistus servus*.

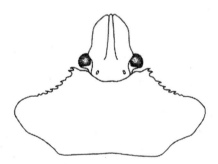

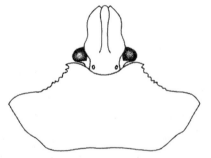

FIGURE 3.37 Head and prothorax, dorsal view, *Euschistus t. luridus*.

FIGURE 3.38 Head and prothorax, dorsal view, *Euschistus t. tristigmus*.

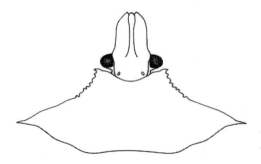

FIGURE 3.39 Head and prothorax, dorsal view, *Euschistus t. tristigmus*.

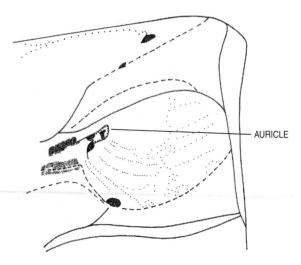

FIGURE 3.40 Metapleuron, *Euschistus t. tristigmus*.

FIGURE 3.41 Beak, lateral view, *Mormidea lugens*.

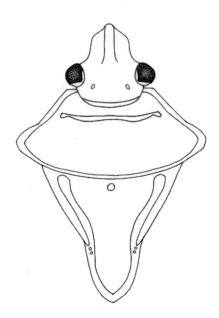

FIGURE 3.42 Head, prothorax, and scutellum, dorsal view, *Mormidea lugens*.

FIGURE 3.43 Head and prothorax, dorsal view, *Proxys punctulatus*.

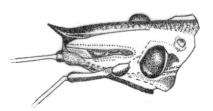

FIGURE 3.44 Head, dorsolateral view, *Proxys punctulatus*.

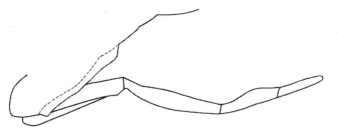

FIGURE 3.45 Beak, lateral view, *Oebalus p. pugnax*.

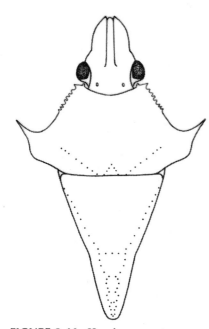

FIGURE 3.46 Head, pronotum, and scutellum, dorsal view, *Oebalus p. pugnax*.

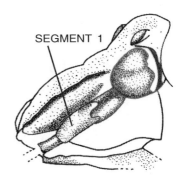

SEGMENT 1

FIGURE 3.47 Head, lateral view, *Oebalus ypsilongriseus*.

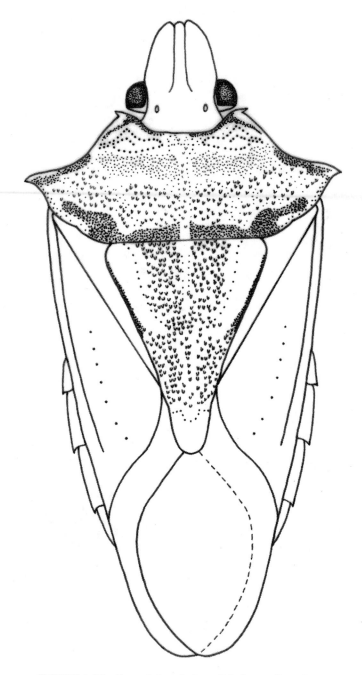

FIGURE 3.48 General dorsal view, *Oebalus ypsilongriseus*.

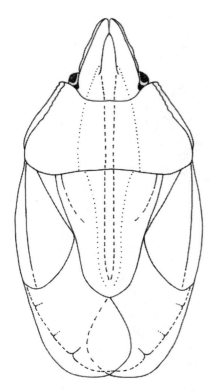

FIGURE 3.49 General dorsal view, *Aelia americana.*

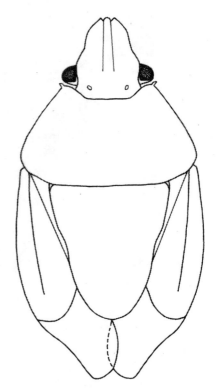

FIGURE 3.50 General dorsal view, *Coenus delius.*

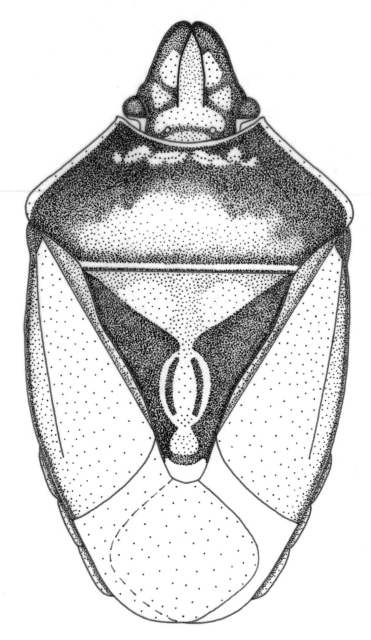

FIGURE 3.51 General dorsal view, *Holcostethus limbolarius*.

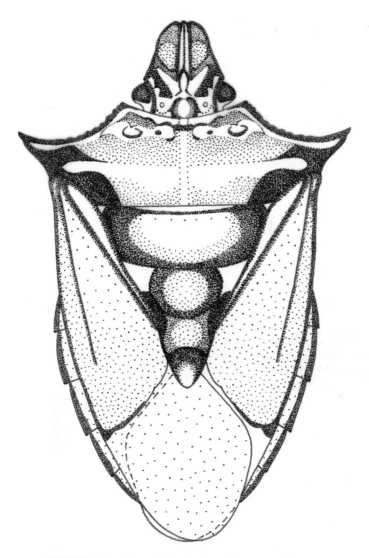

FIGURE 3.52 General dorsal view, *Thyanta perditor*.

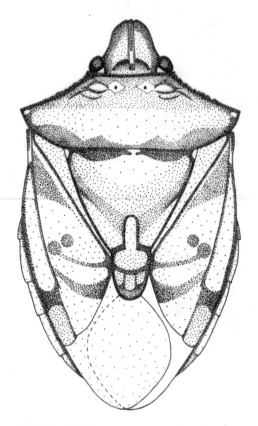

FIGURE 3.53 General dorsal view, *Thyanta calceata.*

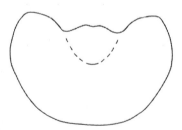

FIGURE 3.54 Pygophore, ventral view, *Thyanta calceata.* (Redrawn from McPherson, J.E., The Pentatomoidea (Hemiptera) of northeastern North America with emphasis on the fauna of Illinois, Southern Illinois University Press, Carbondale and Edwardsville, 1982. With permission.)

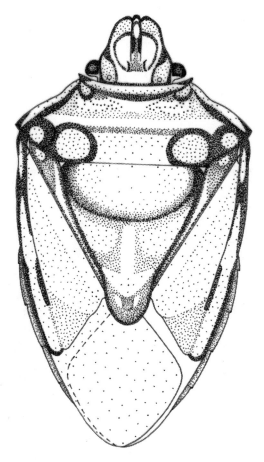

FIGURE 3.55 General dorsal view, *Thyanta c. custator.*

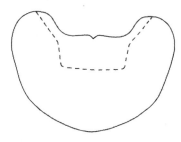

FIGURE 3.56 Pygophore, ventral view, *Thyanta c. custator.* (Redrawn from McPherson, J.E., The Pentatomoidea (Hemiptera) of northeastern North America with emphasis on the fauna of Illinois, Southern Illinois University Press, Carbondale and Edwardsville, 1982. With permission.)

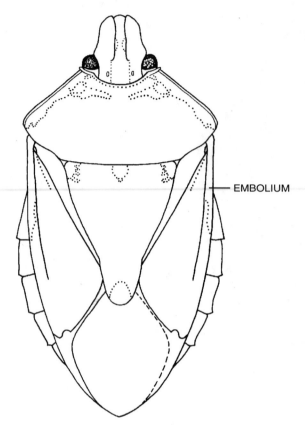

— EMBOLIUM

FIGURE 3.57 General dorsal view, *Chlorochroa sayi*.

FIGURE 3.58 Hemelytron, *Chlorochroa sayi*.

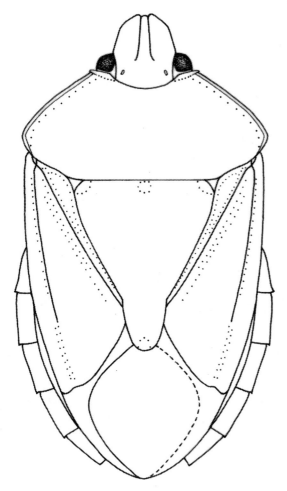

FIGURE 3.59 General dorsal view, *Chlorochroa uhleri.*

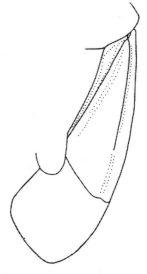

FIGURE 3.60 Hemelytron, *Chlorochroa uhleri.*

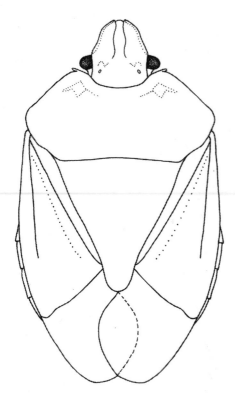

FIGURE 3.61 General dorsal view, *Chlorochroa ligata.*

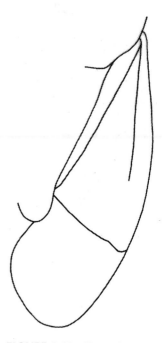

FIGURE 3.62 Hemelytron, *Chlorochroa ligata.*

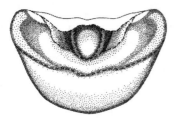

FIGURE 3.63 Pygophore, posteroventral view, *Chlorochroa ligata.*

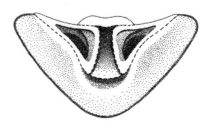

FIGURE 3.64 Pygophore, posteroventral view, *Chlorochroa granulosa.*

4 *Nezara viridula* (L.)

CONTENTS

4.1 INTRODUCTION

Nezara viridula, the southern green stink bug, is cosmopolitan in distribution, occurring throughout the tropical and subtropical regions of Europe, Asia, Australasia, Africa, and the Americas (Todd 1989). Earlier thought to have originated in southern (DeWitt and Godfrey 1972) or southeastern Asia (Menezes et al. 1985a, Todd and Herzog 1980), it is now believed to be of Ethiopian origin (Aldrich 1995a; Aldrich et al. 1989, 1993; Jones 1988; Todd 1989). In North America, it ranges from Virginia and Florida west to Texas and Oklahoma (McPherson et al. 1994) and recently has become established in California (Hoffmann et al. 1987a, McPherson et al. 1994). Other records include Illinois (McPherson and Cuda 1974), Maryland (Aldrich 1990), and New York (Torre-Bueno 1912); the latter two records, and probably all three, are based on adventitious specimens and, in fact, Aldrich suggests the Maryland specimen was blown there from South Carolina by Hurricane Hugo.

Nezara viridula is the most cosmopolitan stink bug pest in North America. It is one of the major stink bug pests of soybean in the world (Kogan and Turnipseed 1987), which is especially true in the United States where it is the most important stink bug attacking soybean and other agricultural crops.

This species is highly polyphagous, attacking over 30 families of dicotyledonous plants and several monocots (Panizzi 2000, Panizzi et al. 2000, Todd 1989). Hoffmann (1935) compiled an early list of the host plants of this stink bug, noting that among the monocots were four species of high economic importance, namely, *Zea mays* L., *Oryza sativa* L., *Triticum vulgare* Villars (*T. aestivum* L.), and *Saccharum officinarum* L. However, he noted that, among the dicots, the bug showed a distinct preference for legumes, an observation that has been supported by subsequent investigators (e.g., Panizzi 2000; Panizzi and Slansky 1991; Panizzi et al. 2000; Todd 1981, 1989; Todd and Herzog 1980). Todd and Herzog (1980) compiled a list of 44 common cultivated and wild host plants for this insect, based on the earlier reports of Jones (1918), Drake (1920), and Hoffmann (1935).

Preference for particular host plants changes with host maturity and phenology (Todd 1981, 1989; Todd and Herzog 1980), with plants being most attractive during fruit and pod formation (Bundy and McPherson 2000b; Chalfant 1973; Lye and Story 1988, 1989; Menezes et al. 1985a; Todd 1981, 1989; Todd and Herzog 1980). As the fruit/seeds mature, the plants become less attractive, and the stink bugs move to more succulent plants (Jones and Sullivan 1982; McPherson and Newsom 1984; Todd 1976, 1981, 1989; Todd and Herzog 1980).

4.2 LIFE HISTORY

The life history of *Nezara viridula* has been documented thoroughly in North America. Two of the earliest studies, Drake (1920) for Florida and Jones (1918) for

Louisiana, were excellent investigations that, even today, are cited frequently although they were conducted before this bug became a major pest.

Nezara viridula has three (Smith et al. 1986) or four generations per year in Louisiana (e.g., Smith et al. 1986, Jones 1918, Todd 1989) and four in Florida (Drake 1920, Todd 1989), and five may occur in southern Florida (Drake 1920, Todd 1989). Three to four generations per year also have been reported outside the United States (e.g., Japan, Kiritani 1964b, Kiritani and Hokyo 1962; Egypt, Ali et al. 1979).

Adults overwinter in protected areas, mainly under litter, bark, and other protected sites (Jones and Sullivan 1981; Smith et al. 1986; Todd 1976, 1989; Watson 1918) including Spanish moss (Rosenfeld 1911, Watson 1918). Jones and Sullivan (1981) found that adults in South Carolina overwintered in above-ground habitats, and none survived the winter when caged over wild radish, deciduous woods litter, or soybean field litter. Ehler (2000) also reported the adults overwintered in above-ground locations, "especially under bark of snag trees in riparian habitats."

Overwintering adults are in reproductive diapause (Harris et al. 1984, Todd 1989) and often can be distinguished by their reddish-brown (or russet) cuticle (Brennan et al. 1977, Harris et al. 1984), although color is not a completely reliable indicator (Seymour and Bowman 1994). Diapause is induced by short-day photophases of 10, 11, and 12 hrs but suppressed by a photophase of 13 hrs (Newsom et al. 1980, Todd 1989). The fifth instars are most responsive to photoperiod influence (Newsom et al. 1980).

Diapausing adults can be active and feed on succculent food during mild periods in the winter (Drake 1920; Newsom et al. 1980; Todd 1976, 1989; Watson 1918); in fact, their survival is enhanced by winter-feeding (Todd 1989). However, few adults remain on succulent plants throughout the winter (Drake 1920, Watson 1918). Mustard (*Brassica* spp.) and wild radish (*Raphanus* sp.) are highly preferred, not only for nourishment but as excellent hibernacula (Todd 1989).

Overwintering survival reportedly is greater for females than for males, for larger than for smaller individuals, and for those with a reddish-brown color (Todd 1989). Overwintering mortality apparently is one of the major limiting factors of population size (Kiritani 1964b, Kiritani and Hoyko 1962, Kiritani et al. 1966a, Todd 1989, Zalom and Zalom 1992). Cold temperature during the winter probably is the most important factor contributing to the annual variation in *Nezara viridula* population densities (Elsey 1993, Jones and Sullivan 1981, Kiritani 1964b, Kiritani et al. 1966a).

Harris et al. (1984) found that, under laboratory conditions, russet-colored (i.e., diapausing) bugs live longer than green bugs, regardless of generation, sex, or reproductive status, and mate and oviposit less. Elsey (1993) found no difference in cold tolerance between diapausing and nondiapausing adults and concluded that diapause primarily serves to regulate the life cycle rather than as a mechanism to survive cold temperatures.

As temperatures increase during early spring, adults leave overwintering sites and begin to feed and mate in clover, small grains, early spring vegetables (e.g., mustard, turnip, beet, radish), corn, tobacco, and weed hosts (e.g., showy crotalaria and coffee senna); females fly up to 1,000 m per day in search of feeding and oviposition sites (Todd 1989). The bugs prefer to feed on plants during fruit formation, with preference varying with the stage of development of their preferred hosts and time of year. After the fruits mature, the plants become unattractive, and the bugs move to more succulent plants.

The resulting nymphs and adults represent the first generation. In temperate regions, midsummer hosts include tomatoes, leguminous weeds, vegetables, row crops, cruciferous vegetation, and okra. By the third generation, soybeans have reached the onset of bloom and podset and now are attractive to adults. Subsequent generations also occur on soybeans. In fact, soybeans are a major source of food in late summer and early fall (i.e., late July through November) (Todd and Herzog 1980). At this time, the bugs apparently reach their highest numbers in soybeans (Buschman et al. 1984, Drees and Rice 1990, Jones and Sullivan 1982, McPherson et al. 1993, Nettles et al. 1970).

Adults and nymphs will attack stems, foliage (particularly leaf veins), flowers, and fruits but prefer young tender growth and, especially, fruiting structures (Todd 1981, 1989; Todd and Herzog 1980). As this bug primarily is a pod, seed, and fruit feeder, its population peak almost always coincides with or lags slightly behind development of the reproductive stages (fruiting) of the primary host species. The sequence of hosts and generations is shown schematically by Todd and Herzog (1980). An additional sequence of hosts is given by Jones and Sullivan (1982).

4.2.1 Mating Behavior and Subsequent Development of Offspring

Precopulatory and copulatory behaviors in *Nezara viridula* have been studied in detail. Mitchell and Mau (1971) reported that males produce a sex pheromone that is attractive to females and to the tachinid parasitoid *Trichopoda pennipes* (Fab.). However, Harris and Todd (1980a), Aldrich (1988a, 1995a), and Aldrich et al. (1987, 1989) found that males and late instars also are strongly attracted to males in the field; Harris and Todd (1980a) noted the attraction occurs throughout the daylight hours but is strongest during the 5-min period just prior to complete darkness. These results, therefore, strongly suggest the olfactory substance acts as an aggregation pheromone rather than as a sex pheromone for *N. viridula* and as a kairomone for *Trichopoda pennipes* (Harris and Todd 1980a). Differences in the chemistry of this aggregation pheromone have been found between geographically isolated populations of this bug (Aldrich 1995a; Aldrich et al. 1987, 1989, 1993).

The entire mating process apparently involves three steps: (1) an aggregation pheromone, which operates over a relatively long distance, is released to bring the sexes together; (2) a sophisticated auditory communication system is utilized in which male and female songs must alternate in a specific order for successful mating to occur; and (3) actual touching behavior occurs (i.e., courtship) (Harris et al. 1982). Further information on the mating process is provided by Borges et al. (1987), Cokl et al. (1972, 1978), and Ryan and Walter (1992).

Touching behavior is initiated by the males who are the aggressors, even though females attain sexual maturity first; males are polygamous and females are polyandrous (Mitchell and Mau 1969). The male approaches the female, places his head beneath her abdomen, and lifts the tip of her abdomen (Mitchell and Mau 1969). If she then lowers her abdomen, the male will repeat the process and will continue to do so until she holds her abdomen in an elevated position. Then the male slides his abdomen along her side while rotating his body until his genital segments come into contact with hers and copulation is initiated. Harris and Todd (1980b) and Ota and Cokl (1991) reported further on precopulatory interaction between the sexes.

A diurnal behavioral catalog of adult *Nezara viridula*, including sexual behavior, has been developed by Lockwood and Story (1986b) and includes 23 behaviors.

Eggs are laid in clusters, closely packed in regular rows, and firmly glued together and to the substrate. The incubation period is 2 to 3 weeks during early spring and fall (Todd 1989, Todd and Herzog 1980) but only 5 (Todd 1989) to 6 (Todd and Herzog 1980) days in the summer.

First instars cluster on or near the egg mass and do not feed (Drake 1920, Todd 1989). They do not use visual cues to aggregate nor do they produce an aggregation pheromone in sufficient quantity until 2 days after emergence; therefore, they apparently use tactile cues during this time to maintain the aggregation (Lockwood and Story 1985a). During the first 2 days, the nymphs produce an increasing amount of what apparently is a bifunctional pheromone (*n*-tridecane), causing aggregation behavior at low concentrations and dispersal at high concentrations (Lockwood and Story 1985a). By the second day, the nymphs have the ability to reaggregate. The first instars also produce (E)-4-oxo-2-decenal, which is not produced by later instars. This compound acts as both an attractant and arrestant on second instars and as a repellent to the fire ant, *Solenopsis geminata* (Fabricius) (Pavis et al. 1994). Aggregation behavior continues through the third stadium and disappears during the fourth (Kiritani 1964a, Todd 1989); in fact, fifth instars may show antagonistic behavior toward each other (Kiritani 1964a). Aggregation behavior seems to speed rate of development (Kiritani 1964a) and reduce mortality (Kiritani 1964a, Kiritani et al. 1966b). Lockwood and Story (1986a) reported that aggregations of first instars survived and developed more rapidly at low relative humidities than isolated nymphs. Also, aggregations accelerated nymphal development at low temperatures and appeared to facilitate intake of atmospheric water and to protect against dessication. And, finally, they reported that aggregated nymphs suffered less predation by the predaceous stink bug *Podisus maculiventris* (Say) and the red imported fire ant, *Solenopsis invicta* Buren.

Nymphs, like adults, are found on those portions of plants upon which they prefer to feed (i.e., tender growing shoots and especially developing fruit and seed). During summer, development from eggs to adults takes 35 to 37 days, depending on temperature (Todd 1989).

4.2.2 Japanese Studies

Various aspects of the biology of *Nezara viridula* have been studied in Japan in such great detail that selected examples will be mentioned. These include

Life cycle
Seasonal sequence of host plants
Mortality factors (i.e., predators, parasites, and weather) (Kiritani and Hokyo 1962)
Effects of biotic and abiotic factors on population size (Kiritani 1964b)

Adult polymorphism (Kiritani 1970)

Changes in the female reproductive system as a predictor of the seasonal life history (Kiritani 1963)

Comparison of fecundity and adult longevity in laboratory and field-cage populations (Kiritani et al. 1963)

Effects of colony size on nymphal developmental rate and survival (Kiritani 1964a)

Movement of females between plants used as food and those used for oviposition (Kiritani et al. 1965)

Effects of temperature on development and mortality (Kariya 1961)

Effects of nymphal and adult densities on fecundity and adult life span (Kiritani and Kimura 1965)

Overwintering survival (Kiritani et al. 1966a)

Factors influencing the upper limits of population density (Kiritani 1965)

Survival and reproduction in relation to population behavior (Kiritani et al. 1966c)

Egg cluster size in relation to the interovipositional period (Kiritani and Hokyo 1965)

Also, a sampling design has been developed for estimating population density (Hokyo and Kiritani 1962). As would be expected, some differences are apparent between those results and those of similar studies in North America, but the overall reports on biology are similar.

4.2.3 NATURAL CONTROL

Jones (1988) compiled an extensive list (from published and unpublished information) of parasitoids that have emerged from *Nezara viridula* worldwide. He reported 57 species among two families of Diptera (i.e., Tachinidae, Sarcophagidae) and five families of Hymenoptera (i.e., Peteromalidae, Eurytomidae, Eupelmidae, Encyrtidae, Scelionidae), 41 of which are egg parasitoids. Many of these parasitoids are associated with this stink bug only incidentally; they are rare or associated more closely with other hosts or habitats. Nymphal stages of stink bugs, including *N. viridula*, generally are not attacked significantly by parasitoids (Jones 1988).

The most successful stink bug parasitoids in the United States appear to be the scelionid egg parasitoid *Trissolcus basalis* (Wollaston) and the tachinid adult and nymphal parasitoid *Trichopoda pennipes* (Awan et al. 1990; Davis 1964, 1966; Jones 1988; Jones et al. 1996; McPherson et al. 1982; Menezes et al. 1985b; Orr et al. 1986). *T. basalis* probably originated in the Old World and now occurs with *Nezara viridula* in the New World (Jones 1988). *T. pennipes* is the most successful adult parasitoid of *N. viridula* in the United States. This species in North America apparently is a complex of cryptic species or biotypes (Jones 1988).

Reports of *Trissolcus basalis* as a major parasitoid of *Nezara viridula* have appeared frequently in the literature. These include, among many others, Buschman and Whitcomb (1980), Mitchell (1965), Nishida (1966), Orr et al. (1986), and Shepard et al. (1994). Females apparently are attracted to *N. viridula* by a component of the bug's defensive secretion from the metathoracic gland; an aldehyde, (E)-2-decenal, appears to act as a long-range kairomone orienting the parasitoid to *N. viridula* populations (Mattiacci et al. 1993). Emergence of *T. basalis* from egg clusters of *N. viridula* exposed to methamidophos (Monitor 4 EC) residue on tomato (*Lycopersicon esculentum* Miller) foliage was not affected, but survival of the emergent adults was reduced (Smilanick et al. 1996).

Trissolcus basalis has been reared successfully in the laboratory under several constant temperatures (Orr et al. 1985a, Powell et al. 1981) and relative humidities (Orr et al. 1985a). Although only applicable indirectly, Orr et al. (1985b) found that *Telenomus chloropus* Thomson, also a scelionid, differed in successful emergence, total fecundity, mean number of progeny, and length of ovipositional period depending upon which of two soybean genotypes (Davis or PI 171444) *Nezara viridula* had been reared.

Powell et al. (1983) developed a stochastic model that simulated interactions of *Trissolcus basalis* with *Nezara viridula* at parasitoid:host ratios of 1:20, 1:60, and 1:100 during a 10-week period (July 9 to September 17) with two releases of the parasitoid. They found the 1:60 ratio would be sufficient to obtain satisfactory control. Orr et al. (1989) studied the effects of permethrin and methyl parathion in soybeans on the survival of *T. basalis*.

The association of *Trichopoda pennipes* with *Nezara viridula* has been known for most of this century. Watson (1918) and Drake (1920) reported *T. pennipes* as a common parasite (= parasitoid) of *N. viridula* in Florida, as did Jones (1918) for Louisiana.

Documentation of the importance of *Trichopoda pennipes* as a natural control agent of *Nezara viridula* continues to appear in the literature. Examples include Buschman and Whitcomb (1980), Eger and Ables (1981), Harris and Todd (1980a, 1981a), McLain et al. (1990a), McPherson et al. (1982), Menezes et al. (1985b), Nishida (1966), Todd and Lewis (1976), and Watson (1934). Temerak and Whitcomb (1984) reported the emergence of one specimen of *T. pennipes* and eight specimens of the sarcophagid *Sarcodexia lambens* (Wiedemann) [as *S. innota* (Walker)] from among 20 adults of *N. viridula* collected in October. The report of *S. lambens* as a parasitoid of *N. viridula* was interesting especially because it had been mentioned previously only by Drake (1920) [as *Sarcophaga sternodontis* (Townsend)].

As noted above, Mitchell and Mau (1971) presented evidence that males of *Nezara viridula* produce a pheromone that not only is attractive to females but to *Trichopoda pennipes*. In addition, they found that males were attacked more frequently than females.

Other investigators, including Todd and Lewis (1976), Menezes et al. (1985b), and McLain et al. (1990a), also have found males of *Nezara viridula* more heavily parasitized by *Trichopoda pennipes* than females. McPherson et al. (1982) reported that males had more eggs deposited on their integument; however, there was no sexual difference in the percentage of bugs with internally developing larvae or in the number of larvae per host. In contrast, Jones et al. (1996) found approximately a 50/50 split in the number of male and female *N. viridula* parasitized by *T. pennipes*. Harris and Todd (1980a) found that *T. pennipes* was most attracted to cages containing *N. viridula* males compared with those containing females or to control cages.

Todd and Lewis (1976) felt that because parasitization of two other species in their study (i.e., *Acrosternum hilare* and *Euschistus servus*) was so low, the male pheromone of *N. viridula*, if present, must be highly specific and not common to all stink bugs. In fact, although the male pheromone of *E. servus* is much different chemically compared with that of *N. viridula*, it is similar to that of *A. hilare* (different ratios) (Aldrich et al. 1989, 1991, 1993). McLain et al. (1990a) felt that because males were more heavily parasitized than females, the risk to females was primarily because of their association with males during courtship and/or copulation. Mitchell and Mau (1971), Harris and Todd (1980a), and Aldrich et al. (1987) reported that more female than male *T. pennipes* were attracted to male *N. viridula* (Mitchell and Mau, Harris and Todd) or to traps baited with male airborne volatiles (Aldrich et al.).

Predation on *Nezara viridula* has been difficult to evaluate because investigators have relied primarily on direct observation. Some of the earliest reports were those of Watson (1918), Jones (1918), and Drake (1920) who reported the following predators as feeding on nymphs and/or adults: *Arilus cristatus* (L.) (Watson, Drake), *Sinea spinipes* (Herrich-Schaeffer), *Zelus cervicalis* Stål (Drake) (Reduviidae); *Euthyrhynchus floridanus* (L.) (Watson, Jones, Drake), *Podisus maculiventris* (Jones, Drake) (Pentatomidae); and *Bicyrtes quadrifasciata* (Say) (Jones, Drake) (Sphecidae).

Stam et al. (1987), using direct field observation and radioactive phosphorus (^{32}P) in a soybean ecosystem, found 18 insect and 6 spider species, representing 6 orders and 19 families, that were predators of *Nezara viridula*. *Solenopsis invicta* (red imported fire ant) appeared to be the dominant egg predator during the vegetative stages of soybean development, but cannibalism of the eggs by nymphs and adults of *N. viridula* frequently was observed. Grasshoppers were more important egg predators during the reproductive stages of soybean development.

Krispyn and Todd (1982) also reported that *Solenopsis invicta* was an important predator of *Nezara viridula*. Adults and nymphs of this bug were caged on soybeans and exposed to this ant. Their numbers were greatly reduced; nymphs were more vulnerable to the ants than adults. Nishida (1966) found the ant *Pheidole megacephala* (Fab.) was an important egg predator of *N. viridula* in Hawaii.

Ragsdale et al. (1981), using an enzyme-linked immunosorbent assay (ELISA), tested 27 species of insects and spiders for both egg and nymphal predation of *Nezara viridula*. They found that two predator complexes were responsible for 91.4 and 95.4% of egg and nymphal predation, respectively. They calculated predator efficiency ratings for the various predators; *Geocoris punctipes* (Say) (Lygaeidae) was one of the top two predators in efficiency in both complexes.

4.2.4 SCENT GLAND SECRETIONS

The odors in pentatomid adults are produced by the metathoracic scent glands (Aldrich 1988a, b, 1995a; Aldrich et al. 1978; Gilby and Waterhouse 1965, 1967; Waterhouse et al. 1961) and dorsal abdominal scent glands (Aldrich 1988b, 1995a; Aldrich et al. 1978, 1995) and in nymphs by the dorsal abdominal scent glands (Aldrich 1988a, b, 1995a; Aldrich et al. 1978; Waterhouse et al. 1961). These secretions apparently have several functions. As examples, they may act as deterrents to predators (Staddon 1979), as sex attractants (Aldrich et al. 1978, Brezot et al. 1994), and as alarm pheromones (Lockwood and Story 1987). In nymphs, they may function both as alarm pheromones (Ishiwatari 1974, Lockwood and Story 1985a) and as aggregation pheromones (Ishiwatari 1974, 1976; Lockwood and Story 1985a), depending on their concentration.

In adult *Nezara viridula*, the secretions of the metathoracic scent gland are comprised of at least 20 components (Gilby and Waterhouse 1965). Two of these components, E-2-hexenal and E-2-hexenyl acetate, significantly increase movement in conspecifics (Lockwood and Story 1987). Interestingly, Aldrich et al. (1995) found that *Trissolcus basalis* females were more likely to attempt oviposition in glass beads coated with an extract from the dorsal abdominal glands of female *N. viridula* than in beads coated only with the solvent; this suggests the wasps were using the odors from these glands as kairomones.

The secretions of nymphal *Nezara viridula* also have been analyzed (e.g., Borges and Aldrich 1992, Lockwood and Story 1985a). A major component is *n*-tridecane. Lockwood and Story (1985a) reported that this compound apparently is a bifunctional pheromone of first instars; it is an attractant at low concentrations and a repellent at high concentrations. Emerging nymphs lack this pheromone and do not reaggregate if separated. However, during the first 2 days, the amount of this pheromone increases as does the ability of these nymphs to reaggregate. Lockwood and Story (1986a) also reported that aggregated nymphs survive better and develop more rapidly at lower relative humidities than isolated nymphs, perhaps because aggregations facilitate intake of atmospheric water and, thus, protect against dessication (Lockwood and Story 1986a). Aggregated nymphs also suffer less predation, perhaps because they are able to pool their exocrine products to confuse or repel predators. Pavis et al. (1994) found that first instars also produce (E)-4-oxo-2-decenal, which is not produced by later instars. This compound acts as both an attractant and arrestant on second instars and as a repellent to the fire ant *Solenopsis geminata*.

4.2.5   Effects of Biotic and Abiotic Factors on Populations

Todd (1989) gives an excellent referenced summary of the effects of biotic and abiotic factors on the population dynamics and mortality of *Nezara viridula*. Factors usually responsible for mortality in the various stages are as follows: eggs are destroyed primarily by parasites and predators; first instars principally by weather (excessive rainfall, low humidity; particularly vulnerable to desiccation during periods of low humidity and high temperatures); second instars mainly by predators (e.g., spiders, ants); third instars to adults because of problems associated with molting; and adults because of parasites.

4.3   LABORATORY INVESTIGATIONS

Nezara viridula has been reared under confined conditions (e.g., Ali and Ewiess 1977; Bowling 1980; Brewer and Jones 1985; Calhoun et al. 1988; Davis 1964, Drake 1920; Ehler 2000; Harris and Todd 1980c, 1981b; Jensen and Gibbens 1973;

Jones 1918; Jones 1985; Jones and Brewer 1987; Kariya 1961; Kiritani 1964a; Kiritani and Kimura 1965; Menusan 1943; Panizzi and Alves 1993; Panizzi and Meneguim 1989; Panizzi and Saraiva 1993; Panizzi and Slansky 1991; Panizzi et al. 1996; Velasco and Walter 1992, 1993; Velez 1974) and the effects of various factors have been studied. Some of these factors have included

Effects of photoperiod (Ali and Ewiess 1977) and temperature (Ali and Eweiss 1977, Kariya 1961) on development and mortality.

Effects of nymphal aggregations on developmental rate and mortality (Kiritani 1964a, Lockwood and Story 1986a).

Effects of natural and artificial diets on developmental rate, nymphal survival (Brewer and Jones 1985, Jensen and Gibbens 1973, Jones and Brewer 1987), and adult longevity (Brewer and Jones 1985, Jones and Brewer 1987).

Effects of seeds and fruits of various plants on survival and developmental time of nymphs and on body weight and size of adults at emergence (Panizzi and Meneguim 1989, Velasco and Walter 1992).

Effects of various legumes on developmental time and on adult body weight, survival, longevity, and reproduction (Panizzi and Slansky 1991).

Differences in survival and reproduction of adults reared on sesame, *Sesamum indicum* L., and soybean (Panizzi and Hirose 1995a).

Differences in survival of starved adults with or without water (Panizzi and Hirose 1995a, b).

Influences of soybean seed protein and oil levels on developmental times and weights of nymphs and adults, and survival to adult (Calhoun et al. 1988).

Effects of the phenological stage of soybean pod development on nymphal survival and stadia, and on adult survivorship, longevity, oviposition, body lipid gain, and body weight (Panizzi and Alves 1993).

Importance of host switching to survival and reproduction (Panizzi and Saraiva 1993, Velasco and Walter 1993).

Nymphal survival appears to be better on maturing seeds than on plants beginning pod formation (Panizzi and Alves 1993). Survival is poor on leaves or stems of legumes, whereas pods and seeds provide a suitable food source (Panizzi and Slansky 1991). Also, within the pod, adults prefer the proximal seed (i.e., nearest the pedicel) over other seeds (Panizzi et al. 1995). Jones (1985) and Harris and Todd (1981b) described rearing procedures (e.g., facilities, equipment, food, rearing cages) for this stink bug.

Several other laboratory-based studies of *Nezara viridula* have been conducted, including

Use of stylet sheaths as an indicator of feeding activity (Bowling 1980).

Sex pheromone communication as influenced by certain physiological conditions (Brennan et al. 1977).

Temporal and numerical patterns of mating and ovipositional behavior (Harris and Todd 1980b).

Fecundity, fertility, and hatch among wild-type and three successive laboratory generations (Harris and Todd 1980d).

Feeding preferences of adults between green and red tomato fruit and between various sizes of green fruit (Lye and Story 1988).

Effects of nymphal and adult densities on fecundity and adult life span (Kiritani and Kimura 1965).

Relationship between egg cluster size and interovipositional period (Kiritani and Hokyo 1965).

Comparison of fecundity and adult longevity in laboratory and field-cage populations (Kiritani et al. 1963).

Adaptive significance of prolonged copulation (McLain 1980).

Relationship of adult size to rate of development, mating success, fecundity, and longevity (McLain 1991).

Correlation between size of once-mated females and their fecundity, the size (i.e., number of eggs) of their largest egg mass, and their adult longevity, and correlation between size of their mates and fecundity (McLain et al. 1990b).

Correlation (or lack thereof) between duration and number of copulations and number of fertilized eggs, and relationship between sperm from two males mating with the same female (sperm precedence) (McLain 1981a).

Heritability of size and response to sexual selection (McLain 1987).

Heritability of male mating success, and correlation between male mating success and female fecundity (McLain and Marsh 1990a).

Apparent ability to adjust sex ratio in response to operational sex ratio (McLain and Marsh 1990b).

Influence of body size of the first male to mate with a female in successfully fertilizing eggs subsequent to the second mating with a different male (McLain 1985).

Preference of a mated female to remate with her former partner or a different male (McLain 1992).

Embryo orientation in egg masses (Lockwood and Story 1986c).

Influence of photoperiod, temperature, and siblings on hatching rhythm (Lockwood and Story 1985b).

Recently, Shearer and Jones (1996a) reported that, under laboratory conditions, *N. viridula* adult females fed for longer durations during the scotophase than the photophase of 24-hour feeding tests. And finally, Mau et al. (1967) and Dyby and Sailer (1999) reported on the effects of ionizing radiation on fecundity and fertility in this stink bug.

4.4 DESCRIPTIONS OF IMMATURE STAGES

The eggs (Figure 4.1) and first to fifth instars have been described (e.g., Drake 1920, Jones 1918).

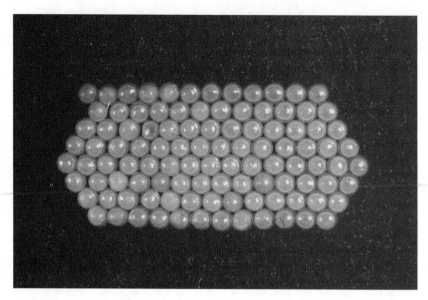

FIGURE 4.1 *Nezara viridula* egg mass. (Courtesy of C.S. Bundy, University of Georgia.)

4.5 ECONOMIC IMPORTANCE

4.5.1 SOYBEAN

4.5.1.1 General Information

Nezara viridula attacks several cultivated crops, but its association with soybean has received the most attention in recent years, this the result of two interrelated factors: (1) increased production of soybean in the southern United States (see earlier discussion of soybean) and (2) the life cycle of this stink bug.

As soybean production has increased, much research has been conducted on increasing yield and quality. This research has covered many areas, one of which has been the thorough documentation of the detrimental effects of feeding by insects including stink bugs.

Several species of stink bugs attack soybean in the United States, the two most important being *Nezara viridula* and *Acrosternum hilare* (Way 1994, Yeargan et al. 1983); *A. hilare* replaces *N. viridula* farther north (McPherson et al. 1994, Turnipseed and Kogan 1976). The two species are joined by *Euschistus servus* in the south as part of a complex (e.g., Clower and Hankins 1960, Boyd et al. 1997). *E. servus*, which is wide-ranging, occurs with *A. hilare* farther north. Other species of some importance as soybean pests, among others, are *Euschistus tristigmus*, *Euschistus variolarius*, *Thyanta* spp., and *Piezodorus guildinii*. *N. viridula* is the most important of these species and notable because it attacks several cultivated crops in addition to soybean.

Prior to soybean expansion, stink bugs were considered minor pests (Newsom et al. 1980). However, following expansion of the crop into the South, these bugs, particularly *Nezara viridula*, adapted particularly well to this new crop. The increase in numbers of *N. viridula* on soybean undoubtedly has influenced its impact on other crops such as cowpea and pecan.

Nezara viridula has three to five generations per year in North America and overwinters as adults in reproductive diapause; however, these adults can become active during milder periods, feeding on succulent plants available at that time. In the spring, as temperatures rise, adults emerge and begin feeding. Populations build up through a succession of hosts, the bugs preferring to feed on plants during fruit formation. After fruits mature, the insects move on to more succulent hosts.

Overwintered adults feed and reproduce on clover, corn, tobacco, and early vegetables (Todd and Herzog 1980). Their offspring represent the first generation, which begins to feed on tomatoes, crucifers, okra, and legumes from April to June. Early planted soybeans are highly attractive to overwintering and first-generation adults (Newsom et al. 1980). Second and third generations occur between May and August, feeding on tomatoes and vegetables. By late July and August, third-generation adults begin migrating into soybean where, subsequently, the fourth and fifth generations are produced (Todd and Herzog 1980). During late summer and early fall, soybean is highly attractive to these bugs (e.g., Spink 1960) and is the primary host (Todd and Herzog 1980). Pecan also is attacked during late summer and early fall, and it and soybean are generally the only two hosts fruiting at this time (Todd and Herzog 1980). However, Jones and Sullivan (1982) stated that soybean and showy crotalaria are the primary late hosts in South Carolina. These late-season crops serve to concentrate the bugs.

The four most common stink bugs on soybean in Georgia are *Nezara viridula*, *Acrosternum hilare*, *Euschistus servus*, and *Piezodorus guildinii*. Their historical first arrivals on this crop in Decatur and Tift counties from 1988 to 1999 are shown in Table 4.1. *N. viridula* and *E. servus* adults were the first species to arrive at both locations each year; their mean Julian days (JD) of arrival, respectively, were 194 and 188 in Decatur County and 184 and 182 in Tift County. The JDs of arrival for *A. hilare* and *P. guildinii* adults, respectively, were 202 and 216 in Decatur County and 207 and 219 in Tift County. *N. viridula* nymphs were the first immature stink bugs observed on average at both sites, JD 216 at Decatur County and JD 213 at Tift County. Most *E. servus* and *A. hilare* nymphs first were observed around JD 220 to 235, whereas *P. guildinii* nymphs first were observed around JD 240 to 260. *P. guildinii* nymphs were not collected in soybean sweep net samples in 6 of the 12 years in Tift County but were not collected in only 1 year (1990) in Decatur County.

4.5.1.2   Damage

More recent reviews on the damage caused by *Nezara viridula* are included in DeWitt and Godfrey (1972, literature survey), Panizzi and Slansky (1985a), Todd (1989), and Todd and Herzog (1980).

Stink bugs, including *Nezara viridula*, damage soybeans directly by their feeding when they pierce the pod and seed to feed on plant juices, simultaneously injecting digestive enzymes (Hammond et al. 1991; Miner 1966a, b; Todd 1976, 1981). Feeding also provides entrance points for disease organisms. Such direct or indirect

TABLE 4.1

Records of First Occurrences of Adult and Nymphal Stages of Four Stink Bug Species Commonly Collected from Sweep Net Samples Taken in Soybeans in Decatur and Tift Counties in Georgia, 1988–1999[a]

	Julian Date When First Stink Bug Was Collected for Each Species (Year)											
	1988	1989	1990	1991	1992	1993	1994	1995	1996	1997	1998	1999
Decatur County, Georgia (Southwest Region)												
1st Sample taken	181	179	171	178[b]	190	181	166	171	184	175	166	182
N.v. adult	194	193	226	—	212	181	180	191	212	175	166	203
N.v. nymph	217	193	226	—	225	238	195	199	219	175	227	266
E.s. adult	194	193	179	—	212	198	180	191	190	183	166	182
E.s. nymph	194	236	233	—	253	—	221	254	247	220	245	266
A.h. adult	201	214	179	—	217	—	207	199	219	210	166	203
A.h. nymph	223	244	233	—	239	189	207	—	226	266	238	—
P.g. adult	217	179	171	—	212	189	195	213	198	175	217	266
P.g. nymph	217	236	—	—	266	238	216	248	263	239	245	266
Tift County, Georgia (South Central Region)												
1st Sample taken	182	164	170	177	181	160	166	179	184	167	177	155
N.v. adult	211	164	186	177	186	160	166	191	212	182	211	155
N.v. nymph	204	221	236	177	216	179	210	226	255	210	225	202
E.s. adult	204	180	179	184	196	160	166	191	191	175	190	165
E.s. nymph	211	243	200	177	230	236	210	226	248	224	225	215
A.h. adult	211	237	207	191	203	195	200	226	227	224	196	165
A.h. nymph	221	180	227	199	203	215	229	233	—	—	237	229
P.g. adult	211	180	—	263	238	202	210	205	—	210	—	256
P.g. nymph	285	—	—	256	238	265	243	255	—	—	—	—

[a] Stink bug species include *Nezara viridula* (N.v.), *Euschistus servus* (E.s.), *Acrosternum hilare* (A.h.), and *Piezodorus guildinii* (P.g.).
[b] Sampling discontinued on Julian day 206 because of inadvertent spray of insecticide.

damage reduces yield and quality. Apparently, the site of the feeding puncture is more important than the number of punctures. A single puncture in the radicle-hypocotyl axis of the seed may prevent germination; several punctures in the coty-ledons may not prevent germination although they may affect vigor (Jenson and Newsom 1972).

It appears that *Nezara viridula* feeds preferentially on soybean seeds in the upper half of the plant until high infestations force the bugs to feed in the lower half; damage-free seeds can compensate for damaged seeds with increases in 100-seed weights of more than 40% (Russin et al. 1987, 1988b). Feeding damage potential is comparable for adults and fifth instars, but less damage is caused by third and fourth instars (McPherson et al. 1979a, Todd and Turnipseed 1974). Interestingly, because of interference or competition, a complex of *N. viridula* and the velvetbean caterpillar, *Anticarsia gemmatalis* Hübner, at certain population densities during R5 and R6, can cause less yield reduction in soybean than either species does alone (Todd and Mullinix 1985).

Significant reductions in yield, quality, and germination can result from feeding by this pest (Duncan and Walker 1968, Jensen and Newsom 1972, Schalk and Fery 1982). Oil and protein levels generally are affected little if at all by moderate to heavy feeding (Miner 1961, 1966a, b; Thomas et al. 1974), whereas the chemical composition of the fatty acids can change (Todd 1976, Todd et al. 1973). However, the oil and protein levels, themselves, can influence developmental times and weights of nymphs and adults (Calhoun et al. 1988). Also, soybean seeds damaged by *Nezara viridula* are more susceptible to attack by stored product pests (Todd 1976, Todd and Womack 1973).

Numerous studies have been conducted to determine the effects of various factors on stink bug feeding and soybean response. These have included the effects of feeding

By various densities of adults (e.g., Cherry 1973; Duncan and Walker 1968; Jones and Sullivan 1983; McPherson et al. 1979a, 1993; Russin et al. 1987, 1988a, b; Thomas et al. 1974; Todd 1976; Todd and Turnipseed 1974) and/or nymphs (McPherson et al. 1979a, Todd and Turnipseed 1974).

On different developmental stages of the seed (Duncan and Walker 1968, Jones and Sullivan 1983, Kilpatrick and Hartwig 1955, Thomas et al. 1974).

On oil and protein (Todd 1976, 1981; Todd et al. 1973).

As related to problems with storage and attack by the cigarette beetle, *Lasioderma serricorne* (Fab.) (Todd 1976, Todd and Womack 1973).

In relation to contamination by microorganisms (Kilpatrick and Hartwig 1955, Ragsdale et al. 1979, Russin et al. 1988a, b).

Damaged seed results in reduced germination, emergence, and seedling survival (Todd 1976, 1981).

Stink bug feeding on soybean reportedly has caused delayed maturity in Louisiana (Boethel et al. 2000), Georgia (Todd and Turnipseed 1974), and Arkansas (Daugherty et al. 1964). The recent study by Boethel et al. (2000) specifically examined the delayed maturity associated with *Nezara viridula* at different population densities exposed to various soybean phenological stages. Their results indicate that delayed maturity occurred only during the pod-set (R3–R4) and pod-filling (R5) stages of plant growth that were exposed to six stink bugs per 0.3 meter of row. Delayed maturity was not as consistent at the late pod-filling stage (R5.5). The greatest reductions in seed yield and quality also occurred when stink bug infestations were present during the R3-R5.5 stages. Thus, if both delayed maturity and yield losses are considered, soybeans need to be protected from injury by *N. viridula* from pod elongation through late pod-filling stages (Boethel et al. 2000).

Jensen and Newsom (1972) developed a system for classifying lightly to heavily damaged seeds, a system still used today, sometimes with some modification. The categories, including the "check" are as follows: none, no visual damage; light, seeds showing punctures but without shriveling of the seed coat; moderate, seeds with some shriveling of the seed coat; and heavy, seeds with extensive shriveling of the seed coat (Figure 4.2). Sometimes a fifth category, severe, is added (e.g., Todd 1976, 1981). The seeds in the heavy and severe categories are of no value for oil, meal, or seed (Todd 1981). Prices paid for damaged seeds may be reduced (Todd 1976, 1981); in fact, a large incidence of stink bug-damaged seeds may even prevent sale of the soybeans (McPherson et al. 1994).

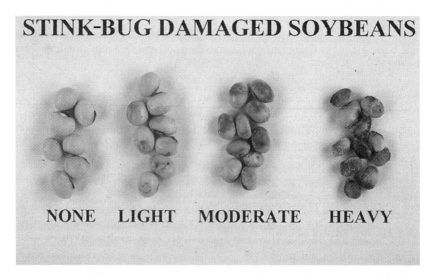

FIGURE 4.2 Soybean kernels exhibiting different categories of stink bug injury. (Courtesy of H. Pilcher, University of Georgia.)

Although several species of stink bugs annually feed on soybean in the southeast (McPherson et al. 1994), the years when stink bugs cause economic losses are years when *Nezara viridula* populations are high (McPherson et al. 1993). The seasonal abundance of stink bugs, including *N. viridula*, varies greatly from year to year in Georgia soybeans (McPherson et al. 1993); estimated losses to the crop ranged from $0 to about $23.5 million in Georgia from 1971 to 1998 (Appendix, Figure A.1).

Jones and Sullivan (1983) and McPherson et al. (1979a) noted that because various stink bug species occur together on soybean, it is advisable that when studying damage, each of the species be included in the field counts.

4.5.2 COWPEA

4.5.2.1 Damage

Nezara viridula is a major pest of cowpea and, as with soybean, prefers to attack the developing seeds. In Georgia, there are two distinct crops of cowpea per year, one that is harvested from June to August, the other from September to October; damage by this stink bug is most serious in the second crop (Nilakhe et al. 1981a). Jones and Sullivan (1982) stated that cowpea is an important host for this stink bug during summer and fall in South Carolina.

The effects on yield and quality resulting from the feeding of *Nezara viridula* on developing seeds have been investigated. For example, Nilakhe et al. (1981a) found that adults and fifth and fourth instars caused more quantitative and qualitative loss of newly emerged pods than did third instars. They also found (1981b) from caged studies that 12 adults per 0.92 m of row could cause significantly greater yield loss than 3 or 6 adults. Loss resulted from reduced seed size and pod length. Twelve adults per 0.92 m of row also resulted in significantly reduced germination.

Fery and Schalk (1981) discovered in outdoor screen cage tests that 0.5 adults per plant at early anthesis caused 94.7 to 96.2% reduction in total seed yield in 1979, whereas 3.0 adults per plant caused 100% loss the same year and 73.9% loss in 1980.

Schalk and Fery (1982) found remarkable differences in several factors resulting from feeding by *Nezara viridula* at early and late bloom. Initial infestation of 3 adults per plant at early bloom resulted in 100% pod damage, a 99% average seed abortion rate, a 100% loss in total seed yield, a 71% average increase in vine weight, and increases of over 100% in number of pod abscission scars on the peduncles; feeding at late bloom resulted in 28% pod damage, a 41% average seed abortion rate, a 74% loss in total seed yield, a more than 100% increase in vine weight, and, again, more than a 100% increase in number of pod abscission scars.

4.5.3 LIMA BEAN

4.5.3.1 Damage

Nezara viridula adults and nymphs feed on lima bean (Figure 4.3) and can cause significant pod drop; adults and fourth and fifth instars generally cause more weight loss in shelled and unshelled pods than third instars (Nilakhe et al. 1981c).

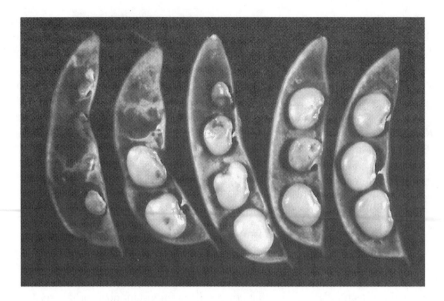

FIGURE 4.3 Lima bean damage in opened pods resulting from stink bug feeding, and undamaged beans on far right. (Courtesy of D.B. Adams, University of Georgia.)

4.5.4 Pecan

4.5.4.1 General Information

The life cycle of *Nezara viridula* as associated with pecan is similar to that associated with other crops. The bug has four or five generations per year and overwinters as adults (Dutcher 1989, Gill 1924, Moznette et al. 1940, Osburn et al. 1954, Payne et al. 1979, Phillips et al. 1964, Polles et al. 1973) under pecan bark, in orchard floor litter (Dutcher 1984, 1990; Dutcher and Todd 1983), and in trash in or near the orchard (Osburn et al. 1954, Payne et al. 1979, Polles et al. 1973); these adults may become active in winter during milder weather (Gill 1924, Moznette et al. 1940, Phillips et al. 1964, Polles et al. 1973). The adults emerge in the spring and lay eggs on the underside of leaves of wild and cultivated crops (Moznette et al. 1940, Osburn et al. 1954, Payne et al. 1979, Phillips et al. 1964, Polles et al. 1973) from April to November (Moznette et al. 1940, Phillips et al. 1964); these plants serve as food for the nymphs (Dutcher 1989, Moznette et al. 1940, Osburn et al. 1954, Payne et al. 1979, Phillips et al. 1964). As the season progresses, the food plants mature and senesce or are harvested. Thus, when the nymphs become adults, they fly to the pecan trees and feed on the developing nuts (Dutcher 1989, Moznette et al. 1940, Osburn et al. 1954, Payne et al. 1979, Phillips et al. 1964). This usually occurs from August through November (Yonce and Mizell 1997). However, drought conditions surrounding the orchard can affect the normal food plants of the bugs, forcing them to search for new hosts any time such conditions occur. Irrigated pecan orchards then become attractive, and the resulting feeding damage can cause extensive nut drop any time during the season (Mizell et al. 1997). Interestingly, the bugs do not breed on this host (Dutcher 1984).

Wood and Tedders (1996) presented a slightly different scenario on the movement of stink bugs into pecan orchards. They stated that the bugs were likely to enter and leave orchards two or more times during the season. When spring weeds or field crops near the orchards begin to mature and senesce, the bugs enter the orchards and feed on the nuts. As the bugs do not reproduce on pecan, they will leave when other plants become available but reenter the orchards when those plants also mature and senesce. They usually penetrate only the first four or five rows, which means spraying can be limited to these same rows. Related to this tendency of the bugs to concentrate along the edges of orchards is the severity of damage a particular orchard may experience from a given population of bugs. Orchards with more perimeter trees, such as those that are long and narrow, are more vulnerable than larger square orchards. Also, smaller orchards near where many of the host plants are beginning to mature or senesce may have higher levels of damage throughout than larger orchards (Wood and Tedders 1996).

4.5.4.2 Damage

Nezara viridula attacks the developing nuts of pecan throughout the growing season, resulting in fruit drop, black pit, and kernel spot. The type of damage is related to the stage of development of the nut when attacked.

Adults feed freely on pecan during nut development. Black pit, which results in premature nut drop, is caused by the bug's feeding on the nuts during the water stage of development (i.e., before the nut shells harden). At this time, the interior of the nut is filled with a watery substance that later will be converted into the kernel (Adair 1932). This damage leads to internal discoloration. Also, the shucks adhere tightly to nuts affected with black pit (Adair 1932). Kernel spot is caused by the bug feeding on the nuts after shell hardening and after the kernel has formed (or is filling). This type of damage generally is not detectable until after harvest when the nuts are shelled. Again, this results in internal damage to the nut and, in fact, can result in a bitter taste (Demaree 1922; Gill 1924; Turner 1918, 1923).

Much of the above information is summarized briefly by Mizell and Tedders (1995), Mizell et al. (1996), Moznette et al. (1940), Osburn et al. (1954), Payne et al. (1979), Phillips et al. (1964), Polles (1977, 1979), Polles et al. (1973), Smith (1996a, b), Stein (1985), and Yates et al. (1991).

Dutcher and Todd (1983) reported that when *Nezara viridula* adults were caged on pecan clusters before shell-hardening (3 adults/cage), they caused 34% fruit drop from mid- to late June and 63% fruit drop from late June to mid-July. Their feeding resulted in black pit and kernel spot, with a mean volume of the kernel spot feeding sites of 5.7 mm^3. The average ingestion of *N. viridula* adults feeding on pecan kernels was 14.1 calories/feeding site compared with that on soybeans of 6.5 calories/feeding site. Obviously, both black pit and kernel spot (together with fruit drop) are of concern to pecan growers.

It now is possible to verify the presence of kernel spot without examining the condition of the kernel. The nuts are stained with a red fluorescent tracing dye, which results in a differential contrast between the shell surface and hemipteran punctures. This technique can be used with a binocular-dissecting microscope to examine the shell exterior for these punctures (Yates et al. 1991).

The characteristics of black pit and kernel spot are given in detail by Adair (1932) and Demaree (1922), respectively.

Payne and Wells (1984) and Reilly and Tedders (1989) discussed microorganisms isolated from lesions in pecan kernels resulting from the feeding punctures of this stink bug.

4.5.4.3   Discovery of Relationship Between *Nezara viridula*, Kernel Spot, and Black Pit

The discovery of both the causal relationship between feeding by *Nezara viridula*, kernel spot, and black pit, and the primary reason for attack of pecan by this bug makes fascinating reading. Kernel spot first was reported as a disease caused by the fungus *Coniothyrium caryogenum* Rand (Rand 1914). However, an understanding of the relationship between kernel spot and the feeding of *N. viridula* did not begin until Turner's preliminary report of 1918. Turner noted that this stink bug was common in middle and southern Georgia and appeared to favor cowpea over all other plants, cultivated or wild. In the fall, when the pea vines dried up, the bugs migrated to nearby plants. Pecan growers commonly sowed cowpea in the groves in early summer so the vines could be turned under later as a "soiling crop." As the vines dried up, usually in September or early October, the bugs would migrate to the pecan. In 1916, there was both a severe infestation of this bug and a severe outbreak of kernel spot. Subsequently, Turner caged bugs on pecan nuts and every nut developed kernel spot. Of several hundred nuts from the same tree not caged, only two or three had spotted kernels. He concluded that it was likely that *N. viridula* was "an important agent in either the actual production or dissemination of this disease."

McMurran and Demaree (1920), cognizant of Turner's (1918) work, reported: "Pecan growers at various points in the South have stated repeatedly to the writers that the appearance of sucking insects during the summer is invariably followed by a more or less seriously affected crop of nuts. The most that can be said at this time is that the disease is a matter for further investigation, and until the cause is definitely determined no recommendations for its control can be made."

Demaree (1922) placed sleeve cages around limbs with immature nuts and added 3 to 7 adults per cage between September 1 and 15 and included control cages (no bugs). He found that kernel spot was present only in nuts caged with bugs. As he could not produce kernel spot with needles, he concluded the damage was caused by the feeding of the bugs. He found a greater prevalence of kernel spot in orchards where cowpea was growing. He noted the common practice of planting cowpea in orchards as a cover crop or hay and that when the pea died down or the hay was harvested, the bugs migrated to pecan nuts, often in great numbers. Also, kernel spot commonly could be found when pecan was grown near tomatoes, okra, or other host plants of this bug. He was not able to produce kernel spots in the laboratory with various inoculations of fungi and bacteria and concluded, therefore, that kernel spot logically resulted from injuries caused by the feeding of sucking insects, including

Nezara viridula. The pathological result of this feeding might be from mechanical rupturing of host cells, by the sucking of plant juices, or by injecting toxic substances into the tissues, or all three. He recommended not using cowpea for a cover crop in pecan orchards but, rather, substituting some other legume such as velvet bean, which he noted was not favored by the bug.

Turner (1923) conducted a study in Georgia similar to that of Demaree (1922) except that his study was conducted over an extended period. Beginning on August 21, he placed 5 *Nezara viridula* adults in each of 10 cages enclosing nuts for approximately 10 days, then removed the bugs, placed other bugs (5 adults, presumably) in 10 more cages, and continued this procedure until harvest (this included six 10-day series ending on October 19); the results were compared with the check (50 cages). The bugs in the first series (August 21 to September 1) caused all the nuts to drop, those in the second series (September 1 to 11) caused 72% to drop, and those in the remaining series caused no drop. All nuts exposed to the bugs in the second, third, and fourth series (September 1 to 11, September 11 to 20, and September 20 to October 4) had kernel spot. The percentage of nuts with kernel spot dropped to 50% during the fifth series (October 4 to 12), and to almost zero in the sixth series (October 12 to 19). In comparison, none of the nuts in the check cages dropped or had kernel spot. He noted that adults fed freely on the developing nuts up to harvest and were able to pierce the shells, even when the shells were hard. He found that only adults caused kernel spot because nymphs could not survive on the nuts. This indicated, then, that pecan was a feeding host and not a breeding host. He stated that the bug reproduced freely on legumes including cowpea, soybean, and mung beans.

Gill (1924) reported that although *Nezara viridula* could be common on pecan, rarely were nymphs found; this indicated, then, that the bug did not reproduce on pecan and, thus, supported Turner's (1923) conclusion. Gill stated that cowpea and soybean should not be planted in pecan orchards and, as did Demaree (1922), recommended that velvet bean be substituted.

Black pit was listed by McMurran and Demaree (1920) under their section "Diseases due to unknown causes." They described the symptoms of black pit and stated, "It remains for further investigation to throw light on this very obscure disease."

Adair (1927, 1932) investigated black pit in a manner similar to that used for kernel spot. He (1927) reported that pecans punctured with an insect pin developed the characteristic color of black pit. He then enclosed clusters of pecans in sleeve cages, added *Euschistus servus* or *Leptoglossus oppositus* (Say) to some, and used the other cages as checks. He found that only those cages containing bugs developed black pit. Subsequently (1932), he reported on a similar (same?) caged experiment in more detail. For this experiment, he placed sleeve cages over limbs containing nuts and added 10 to 15 specimens of *E. servus*, *L. oppositus*, or *L. phyllopus*. As with Demaree's (1922) work, only those cages with bugs subsequently showed black pit damage. Although Adair did not test *Nezara viridula*, he stated that it was known to puncture pecans in feeding and cause black pit. He suggested using the same cultural control methods used for controlling kernel spot (i.e., substituting velvet bean for cowpea or soybean as a summer cover crop because velvet bean was not a preferred host). Also, he recommended destroying all native host plants (e.g., purple-flowered thistle, basketflower, yucca, jimson weed) of this bug in the spring that were growing in or near pecan orchards and not planting preferred cultivated crops (e.g., cowpea, bean, squash, tomato) in or near pecan.

4.5.5 MACADAMIA

4.5.5.1 General Information

Similar to the relationship between *Nezara viridula* and pecan is that between this bug and macadamia, where feeding damage to the nuts also has been reported. Kernel spotting first was observed on October 8, 1962, in nuts harvested in Nuuanu Valley, Oahu, Hawaii (Mitchell et al. 1965). Mitchell et al. (1965) noted the damage appeared similar to kernel spot of pecans and that *N. viridula* was known to cause this damage by its feeding. Adults and second to fifth instars of this same bug also fed on macadamia nuts. Although all stages had been observed on macadamia, only one egg cluster had been found, indicating that this bug preferred to lays its eggs elsewhere. Damage usually was not apparent until after shelling. The bugs could penetrate the husk and shell of a mature nut to reach the kernel, with a single kernel having as many as 70 or more feeding punctures. They examined several orchards during the harvest season (August through January) for damaged kernels resulting from feeding by this stink bug. The orchard with the most damage was one in which the ground beneath the trees was bare, but the orchard, itself, was surrounded by experimental corn, tomato, alfalfa, and bean plantings. The plantings produced many stink bugs and, when these hosts were harvested, the bugs moved to other areas including macadamia trees where they began feeding on the nuts. The highest mean kernel damage from this feeding was 56.2 and 7.2% for 1963 and 1964, respectively.

Recently, Shearer and Jones (1996b) reported that macadamia nut was not a suitable host plant for *Nezara viridula*, thus supporting Mitchell et al.'s (1965) suggestion that this bug preferred to lay its eggs elsewhere. They found that weight gain, fecundity, and survival of females fed macadamia nuts were inferior to females fed a standard diet of green beans and peanuts. Also, no second or third instars and only a few fourth and fifth instars became adults when they fed only on macadamia nuts. Nymphs that did survive on the nuts weighed less and took longer to reach adulthood than those feeding on the standard diet. Thus, *N. viridula* populations apparently do not increase on macadamia in the field and need other host plants to survive and reproduce.

4.5.5.2 Damage

Feeding by *Nezara viridula* on macadamia nuts causes discolored spotting and pitting on the kernels (Mitchell et al. 1965, Oi 1991); this results in downgrading of the nuts from this damage or from the introduction of contaminants and from increased processing costs (Oi 1991). Feeding also results in a significant increase in nut abortion rates for smaller nuts (10 to 28 mm diam.) but not for larger nuts (Jones and Caprio 1994). The bugs not only attack the full-sized nuts when they are on the tree but, also, after they fall to the ground, generally within 1 week. As the magnitude of damage to nuts on the ground appears higher than on the trees, pest management programs should include practices that reduce both canopy and ground populations of the bugs (Jones and Caprio 1994).

4.5.6 RICE

4.5.6.1 Damage

Feeding by *Nezara viridula* on rice has been reported from Florida (e.g., Genung et al. 1979, Green et al. 1954) and Louisiana (Gifford et al. 1968a, Smith et al. 1986). Gifford et al. (1968a) and Smith et al. (1986) noted that it damages rice in Louisiana only on occasion but particularly following mild winters when large populations have built up on adjacent soybean and have overwintered successfully. Smith et al. (1986) further stated that the bugs feed on stems of young rice plants just before head emergence from the boot. Feeding causes the emerging heads to be discolored, curved, and partially sterile; the discoloration consists of a light-green, dark-green mosaic appearance of the leaves. Generally, the bugs do not feed on the panicles. In fact, Gifford et al. (1968a) reported the bugs could survive and reproduce on vegetative rice in the greenhouse.

4.5.7 WHEAT

4.5.7.1 Damage

Nezara viridula can damage wheat kernels, the severity of which is related directly to the density of the bugs and, therefore, the amount of feeding (as determined by feeding sheaths) (Viator and Smith 1980; Viator et al. 1982, 1983). Viator and Smith (1980) studied the effects of its feeding on the developing heads at various infestation levels (0, 1, 3, and 6 sexed pairs of adults/15 caged heads); the insects were caged at heading and remained until the plants were harvested. The association between the bugs, subsequent germination, and physical damage was determined by calculating correlation coefficients. They found that seed quality, as measured by germination and damage, was affected adversely by the feeding of the bugs.

Viator et al. (1982, 1983) followed up the above experiment with a more detailed study, using similar infestation levels (0, 1, 3, and 6 sexed pairs of adults/20 caged heads) but, specifically, at the milk and soft-dough stages of grain development. They found that at each infestation level, there was a decrease in germination percentage, kernel weight, and baking quality as the number of stylet sheaths increased. The bugs fed more on kernels infested at the milk stage than at the dough stage. When kernels were attacked in the milk stage by as few as 1 pair of bugs, there was a significant reduction in germination, kernel weight, and kernel texture. Conversely, when the heads were attacked by 1 pair of bugs in the soft-dough stage, there was no significant reduction in these kernel characteristics. They noted that infestation levels rarely exceed 2 adults/20 heads in the field and, therefore, control measures are warranted only for wheat infested in the milk stage. Incidentally, they noted (1983) that the bugs produced approximately twice as many sheaths on the heads in the milk stage than in the soft-dough stage. Finally, they noted (1983) that because the damage caused by *Nezara viridula* and *Oebalus p. pugnax* is virtually indistinguishable, the two taxa should be grouped together when control recommendations are being developed.

4.5.8 SORGHUM

4.5.8.1 Damage

Nezara viridula attacks sorghum (e.g., Cronholm et al. 1998; Fuchs et al. 1988; Hall and Teetes 1980, 1981, 1982b, c; Hall et al. 1983; Wiseman and McMillian 1971) but, in Texas, is not present on the plants until the last half of July when the grain is in the soft and hard-dough stages of development (Hall and Teetes 1981). It will attack panicle stems, rachis branches, and glumes but is primarily a seed feeder where it causes reduction in yield and germination (Hall and Teetes 1980, 1982b). Sorghum is most susceptible to damage when it is attacked in the milk stage, and the amount of damage is directly related to the density of the bugs (Fuchs et al. 1988; Hall and Teetes 1980, 1982b, c). The damage threshold is approximately 4 bugs per panicle from the milk stage to maturity (Hall and Teetes 1980, 1982b).

Hall and Teetes (1981) listed several alternate host plants of *Nezara viridula* found near sorghum in central Texas that could serve as potential sources of bugs prior to and during sorghum grain formation.

4.5.9 CORN

4.5.9.1 Damage

Nezara viridula attacks corn (Clower 1958, Negron and Riley 1987, Riley et al. 1987) and is capable of causing significant damage, particularly at the V15 stage (top ear, 1.5 to 2.0 cm long) (Negron and Riley 1987, Riley et al. 1987). The damage threshold is 2 adults per plant until V15 when the bugs apparently are most damaging (Negron and Riley 1987, Riley et al. 1987). Damage at V15 ranges from ears that are destroyed totally to those varying in length reduction and kernel filling (Riley et al. 1987). Feeding damage is most severe in younger corn plants, and yield reductions result from total ear loss rather than reduction in kernel weight (Negron and Riley 1987, Riley et al. 1987). Riley et al. (1987) found no effect of these bugs on germination and seed weight, at least at the adult densities tested (0, 2, 4, and 6 bugs/plant). The bugs usually are found in the margins of the field but may spread into adjacent areas when they become abundant (Riley et al. 1987). Estimated losses to the crop in Georgia from stink bugs, including *N. viridula*, ranged from $0 to almost $11 million from 1971 to 1997 (Appendix, Figure A.2).

4.5.10 TOMATO

4.5.10.1 Damage

Nezara viridula will attack tomatoes (e.g., Anonymous 1998; Callahan et al. 1960; Chalfant 1973; Hoffmann et al. 1987a, b, 1991; Kelsheimer and Wolfenbarger 1952; Lye and Story 1988, 1989; Lye et al. 1988a, b; Swan and Papp 1972; Zalom and

Zalom 1992). It shows a preference for fruit clusters over vegetative structures (Lye and Story 1989). It prefers mature green tomatoes over mature red tomatoes if the fruits are of equal size and feeds less frequently on intermediate size tomatoes if the fruits are green (Lye and Story 1988). Feeding induces early maturity of the fruit, resulting in reduced size and weight (Lye et al. 1988a), a bitter taste, and pithy texture (Callahan et al. 1960). Also, the grade of the tomato is affected significantly by bug density and feeding duration (Lye et al. 1988b).

In Georgia, from 17 to 22% of the untreated tomatoes have been damaged by feeding of *Nezara viridula*, with 1 to 2 feeding spots per fruit; however, only 0 to 8% of tomatoes treated with effective insecticides (acephate and monocrotophos) have been damaged, with 0 to 0.02 feeding spots per fruit (Chalfant 1973). Although introduced into California in the 1980s, *N. viridula* populations have remained relatively low on tomatoes, probably due to cold winters and *Trissolcus basalis* egg parasitism (Zalom and Zalom 1992).

4.5.11 TOBACCO

4.5.11.1 Damage

Nezara viridula, in earlier literature, was listed as occurring on tobacco but seldom as important (e.g., Rabb et al. 1959). However, today it is reported to be a pest of flue-cured tobacco (Reich 1991). Both adults and nymphs extract plant fluids from young tender growth and may cause the leaves to wilt and flop over due to the loss of turgor pressure (Figures 4.4 and 4.5). Generally, the plants recover from this injury, even at maintained population densities of 5 adults per plant (R.M. McPherson, unpublished data), although some leaves dry up similarly to sunscald.

4.5.12 COTTON

4.5.12.1 Damage

Nezara viridula occasionally has been listed from cotton (Morrill 1910, Swan and Papp 1972), but the limited information associated with these earlier reports suggests this stink bug is not of economic concern on this crop. However, its economic potential may need reassessment because it now appears it is capable of causing economic damage (Figures 4.6 to 4.9). Turnipseed et al. (1995) recently have reported that during their study of *Bacillus thuringiensis* (Bt) cotton in South Carolina, *N. viridula, Acrosternum hilare,* and *Euschistus servus* were the major secondary pests observed in untreated Bt cotton. They found 7% damaged bolls in non-Bt cotton treated twice with Karate® insecticide compared with 27% damaged bolls in untreated Bt cotton. Recently, Greene et al. (1999) investigated the effects of feeding by adults and nymphs on the bolls of Bt cotton in field-cage studies in Blackville, SC, and Tifton, GA. To determine the extent of feeding damage by various instars (second to fifth) and the adults, bugs were confined individually on 8- or 9-day-old bolls (days from white bloom) for 3.5 days and the bolls examined 4 days later for damage as evidenced by wartlike growths on the interior carpel walls. To determine the effect of boll age on the severity of damage, fifth instars were confined individually on 4-, 8-, 14- or 15-, and 18 day-old bolls for 5 days and the bolls examined 4 days later for feeding damage. Finally, to relate the extent of boll damage to individual boll yield, young fifth instars were confined individually to 13-day-old bolls for 7 days and the bolls examined 4 days later for internal growths or allowed to mature. Yield was determined by weighing all seed cotton from open and partially open locks. Controls were used in all experiments. The authors reported that fifth instars caused more damage than adults or younger nymphs. Boll damage by fifth instars decreased as bolls matured, and yield from bolls exposed to fifth instars was reduced by 59% compared with unexposed bolls. Bundy et al. (2000) also reported on boll damage. They found that the presence of a feeding stylet sheath (Figure 4.7) was the first indicator of internal boll damage (Figures 4.8 and 4.9).

Certainly, this is limited information, but it does show that this highly polyphagous species has the potential to be a significant pest on this crop. In fact, Hollis (2000) reported that stink bug pests, all species combined, ranked as the fifth most costly pest in the Cotton Belt during the 1999 season, according to figures presented by Michael Williams, Mississippi State University, Mississippi State, at the 2000 Beltwide Cotton Conference. Approximately 25% of the entire cotton acreage was infested with stink bugs, reducing yields 0.4%, which accounted for losses of more than 97,000 bales (representing more than $66 million). All of the cotton-producing states except Virginia, Arkansas, Oklahoma, and California reported that stink bugs were pests of the 1999 crop. Estimated losses to the crop in Georgia from stink bugs, including *N. viridula*, ranged from $0 to about $11.5 million from 1971 to 1978 (Appendix, Figure A.3).

4.5.13   OTHER CROPS

4.5.13.1   Damage

Nezara viridula has been reported from oats, rye (Anonymous 1974), peach (Snapp 1941, 1954; Swan and Papp 1972), tangerine, oranges (Watson 1918), potato (Morrill 1910, Radcliffe et al. 1991, Swan and Papp 1972), turnip (Morrill 1910), okra, and other vegetables (Swan and Papp 1972). Recently, it has been observed feeding on peanuts and causing leaves to flop from loss of tugor pressure (much like in tobacco), although it is believed that this plant is an interim host and sustains little or no damage (Jim Todd, personal communication).

FIGURE 4.4 Tobacco plant damage resulting from stink bug feeding (wilting or flopping leaf). (Courtesy of D.C. Jones, University of Georgia.)

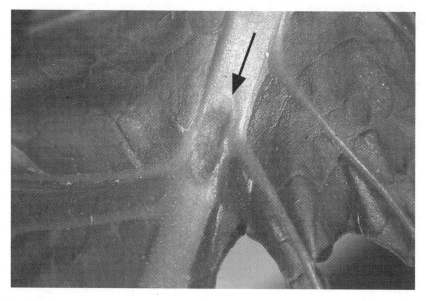

FIGURE 4.5 Stink bug feeding site on tobacco leaf vein. (Courtesy of C.S. Bundy, University of Georgia.)

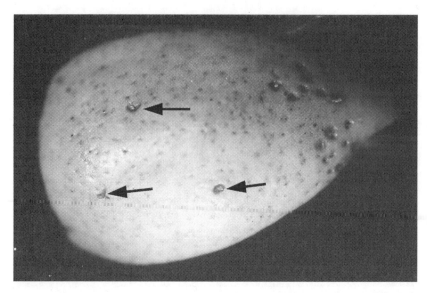

FIGURE 4.6 External cotton boll damage resulting from stink bug feeding. (Courtesy of C.S. Bundy, University of Georgia.)

FIGURE 4.7 Stink bug stylet sheath (40×) remaining on cotton boll after feeding. (Courtesy of C.S. Bundy, University of Georgia.)

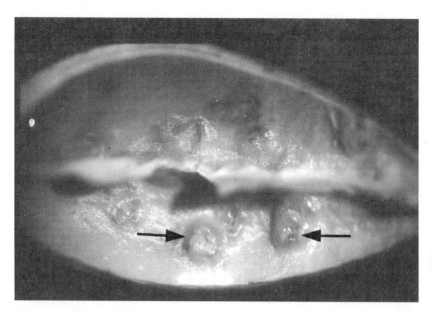

FIGURE 4.8 Internal warts on cotton boll 24 hours after exposure to feeding by stink bugs. Undamaged boll has smooth internal surface. (Courtesy of C.S. Bundy, University of Georgia.)

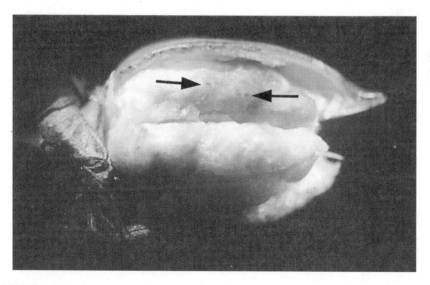

FIGURE 4.9 Internal cotton seed damage (see arrows) and surrounding cotton lint staining 6 days after initial feeding exposure to stink bugs. (Courtesy of C.S. Bundy, University of Georgia.)

5 *Euschistus* spp.

CONTENTS

5.1   INTRODUCTION

Several species of *Euschistus* occur in North America (Froeschner 1988). However, based on the literature, only *E. servus*, *E. variolarius*, *E. tristigmus*, and *E. conspersus* have been of continuing economic importance, particularly *E. servus*. *Euschistus* spp. often are part of a stink bug complex attacking various cultivated crops (e.g., Jones and Sullivan 1983, McPherson et al. 1979a).

Euschistus spp. are polyphagous, feeding on grasses, shrubs, and trees. As with phytophagous bugs in general, they prefer those plants that are producing fruits and pods. As such, the above four species seriously can affect quality and yield of cultivated crops, from agricultural crops such as soybean, wheat, alfalfa, and corn to horticultural crops such as tomato, peach, pear, apple, and pecan (McPherson 1982).

The annual life cycles of the *Euschistus* spp. considered here are similar, with the possible exception of *E. variolarius*. All overwinter as adults in leaf litter, crop residue, or any other convenient shelter, and most species are bivoltine throughout much of their ranges (they may be univoltine in the northern United States and Canada). The primary exception is *E. variolarius*, which has been listed as both uni- and bivoltine (McPherson 1982).

5.2   SCENT GLAND SECRETIONS

The taxonomic similarity of *Euschistus* spp. is exemplified by some recent work of Aldrich (1996) and Aldrich et al. (1991) on adult pheromone composition. They isolated and identified methyl (2E,4Z)-decadienoate as the major male-specific volatile of *E. servus*, *E. tristigmus*, *E. conspersus*, *E. politus* Uhler, and *E. ictericus* (L.). In field tests conducted in Maryland and California, males, females, and nymphs of *E. servus*, *E. tristigmus*, *E. conspersus*, and *E. politus* were attracted to this ester. Also, in Maryland, tachinid parasitoids [*Gymnosoma* spp., *Euthera* spp., *Gymnoclytia occidua* (Walker), and *Euclytia flava* (Townsend)] were attracted to this volatile and, therefore, were using it as a host-finding kairomone. In a sixth species, *E. obscurus* (Palisot de Beauvois), which extends from the Greater Antilles and Mexico into the southern United States (Froeschner 1988), this volatile was one of two that were produced in abundance by males; the other was tentatively identified as methyl 2,6-dimethyltetradecanoate (Aldrich et al. 1991), later corrected to methyl 2,6,10-trimethyltridecanoate (Aldrich 1995a, 1996; Aldrich et al. 1994).

Aldrich et al. (1995) analyzed the dorsal abdominal gland secretions of the adults of several stink bug species including *Euschistus servus*, *E. tristrigmus*, and *E. politus*. Six-carbon compounds, specifically hexanal, (E)-2-hexenal, and hexanol were most abundant among the three *Euschistus* species of the 15 volatile compounds identified, but the relative percentages differed. Also, the various volatiles were not shared by all three species. At the concentrations tested, they concluded these secretions might be epideictic, promoting spacing in the natural environment.

The nymphal secretions of *Euschistus tristigmus* have been analyzed further by Borges and Aldrich (1992).

5.3   *EUSCHISTUS SERVUS* (SAY)

5.3.1   INTRODUCTION

Euschistus servus, the brown stink bug, occurs throughout North America and is comprised of two subspecies, *E. s. servus* (Say), which occurs from the southeastern United States west through Louisiana, Texas, New Mexico, and Arizona into California; and *E. s. euschistoides* (Vollenhoven), which occurs across Canada and the northern part of the United States (McPherson 1982). The two subspecies intergrade in a broad band from Maryland to Kansas (Sailer 1954). *E. impictiventris* Stål is a synonym of *E. servus* (Rolston 1974).

Euschistus servus is a highly polyphagous pest, feeding on grasses, shrubs, and trees (see McPherson 1982). It attacks both uncultivated and cultivated plants and, in fact, its ability to survive on weedy hosts probably plays a major role in its pest status on many cultivated crops such as alfalfa, corn, cotton, fruits, and soybean. Interestingly, it occasionally has been reported as predaceous; records include the cotton worm [*Alabama argillacea* (Hübner)] (Riley 1885); mountain-ash sawfly, *Pristiphora geniculata* (Hartig) (Beaulne 1939, Forbes and Daviault 1964); and imported cabbageworm, *Pieris rapae* (L.) (Culliney 1985).

5.3.2   LIFE HISTORY

Euschistus servus is bivoltine and overwinters as adults (e.g., Adair 1932; Forbes 1905; McPherson and Mohlenbrock 1976; Munyaneza and McPherson 1994; Phillips and Howell 1980; Porter et al. 1928; Rolston and Kendrick 1961; Woodside 1946a, 1947, 1950) under crop residues (Jones and Sullivan 1981), leaves, pieces of bark and similar objects, and in bunches of grass (Adair 1932). It prefers to overwinter in open fields compared with deciduous or pine woodlands or in the field–woodland edge (Jones and Sullivan 1981).

Adults emerge during spring (March and April) and frequently are found on mullein (e.g., Munyaneza and McPherson 1994; Rolston and Kendrick 1961). In southern Illinois, first (summer) generation nymphs are found from May to August, and second (overwintering) generation nymphs from August to late October (Munyaneza and McPherson 1994).

5.3.2.1   Mating Behavior

Precopulatory and copulatory behavior has been studied in this species (Drickamer and McPherson 1992, Youther and McPherson 1975). Touching is initiated by the male who begins by antennating the female's antennae, head, and pronotum. Then, he moves posteriorly, antennating her thorax and abdomen. At this point, and while antennating the underside of her abdomen, he attempts to lift her posterior end with his head. If she is receptive, she holds her abdomen in an elevated position. The male then turns 180° and, with his abdomen elevated to approximately the same angle as hers, and, with his aedeagus extended, backs toward her until their genital segments come into contact. He then moves his abdomen from side to side during insertion. During copulation, the male rests his hind legs on hers, their bodies forming approximately a 40° angle with the substrate. An unreceptive female will not raise her abdomen and often will kick at the courting male.

5.3.3 LABORATORY INVESTIGATIONS

Euschistus servus has been reared under confined conditions (e.g., Brewer and Jones 1985, Esselbaugh 1948, Foot and Strobell 1914, Munyaneza and McPherson 1994, Rolston and Kendrick 1961, Russell 1952, Sailer 1952, Woodside 1946a), and its fecundity and fertility (Youther and McPherson 1975) have been studied.

5.3.4 DESCRIPTIONS OF IMMATURE STAGES

The eggs (Deitz et al. 1976, Esselbaugh 1946, Javahery 1994, Munyaneza and McPherson 1994) and various instars (DeCoursey and Esselbaugh 1962, Morrill 1910, Munyaneza and McPherson 1994) have been described. Munyaneza and McPherson's study also included comparisons of the egg and first to fifth instars of *Euschistus servus* with the corresponding stages of *E. variolarius*, which they closely resemble.

5.3.5 ECONOMIC IMPORTANCE

5.3.5.1 Soybean

5.3.5.1.1 General Information
Soybean is attacked by a complex of stink bugs in the continental United States. The species with the greatest economic impact is *Nezara viridula* (Panizzi and Slansky 1985a). This species generally occurs only in the southern states where it is associated most often with *Acrosternum hilare* and *Euschistus servus* (e.g., McPherson and Newsom 1984; McPherson et al. 1994; Russin et al. 1987, 1988a, b; Todd 1976); however, *N. viridula* usually comprises the major proportion of samples involving these three species (e.g., Bundy and McPherson 2000b; Jones and Sullivan 1979, 1983; McPherson and Newsom 1984; Russin et al. 1987, 1988a, b). Farther north (including as far south as Kentucky and parts of Arkansas), *N. viridula* is replaced by *A. hilare* (McPherson et al. 1994, Miner 1966a, Turnipseed and Kogan 1976), which is still associated with *E. servus* (e.g., Miner 1966a, b; Tugwell et al. 1973). Zeiss and Klubertanz (1994) listed *N. viridula* and *A. hilare* as major pests of soybean and *E. servus* as a minor pest.

Occasionally, other species can become economically important on soybean including *Euschistus tristigmus* (Jones and Sullivan 1979, McPherson and Newsom 1984, McPherson et al. 1994), *Euschistus variolarius* (McPherson et al. 1994),

Thyanta custator accerra (Jones and Sullivan 1982), and *Piezodorus guildinii* (McPherson et al. 1993, 1994).

5.3.5.1.2 Damage

Euschistus servus potentially is capable of causing severe damage to soybean. Jones and Sullivan (1983), using 1 or 3 caged adults per 0.46 m (18 in) of row of soybean during the R4 and R5 plant growth stages, found no difference in the type and amount of feeding damage between *E. servus*, *Acrosternum hilare*, and *Nezara viridula*. Russin et al. (1988b) also stated that these same species were approximately equal in their ability to damage soybeans.

Several investigators have reported on the types of damage caused by *Euschistus servus*. Daugherty et al. (1964), using field cages containing 6 soybean plants and densities of 1, 2, 3, or 4 pairs of adults per cage, found that as infestation levels increased, the number of undeveloped beans was most pronounced in the middle one-third of the plant. At densities of 3 or 4 pairs, plant maturity was delayed and the pods continued to develop on the upper part of the plant beyond the time such growth usually ended; these same plants remained green and actively grew well beyond the time plants in adjacent fields and the other cages matured. They found a correlation between the number of damaged beans and the number of insects present. Also, as the number of feeding punctures increased, the oil content of the beans was reduced and protein content increased. McPherson et al. (1979a) reported that, under field-cage conditions, fourth and fifth instars and adults could cause as much or more damage to soybeans when compared with fourth instars of *Nezara viridula*. Bickenstaff and Huggans (1962), using a mixture of *E. servus* and *Euschistus variolarius* in field cages containing plants at various stages of development beginning at full bloom, found the plants showed progressively less damage as they matured from full bloom or just past bloom to leaf drop. Finally, Daugherty et al. (1964) found that as feeding punctures increased, germination of the beans decreased; they suspected this was the result of the introduction of certain unidentified organisms.

Daugherty (1967) reported that under caged conditions, *Euschistus servus* was capable of transmitting the causative agent of yeast-spot disease, *Nematospora coryli* Peglion, of soybean. Later, Russin et al. (1988a, b) reported that stink bug feeding could increase the incidence of certain seedborne fungi and bacteria as levels of stink bug infestation and damage increased.

Euschistus servus is one of the most common stink bugs in North America and, potentially, should be a major pest of soybean. Furthermore, its populations are highest in the fall (e.g., Jones and Sullivan 1983, McPherson et al. 1993) when soybean often is infested severely with stink bugs. Yet, its numbers on soybean are generally low compared with those of *Nezara viridula*. So, why is it not more of a problem on soybean than it is? The answer may lie in the results of Jones and Sullivan's (1982) investigation in South Carolina. They studied the seasonal abundance of *E. servus* adults and nymphs on uncultivated and cultivated plants (including soybean). They found the most important spring hosts were pepper grass (*Lepidium virginicum* L.), vetch (*Vicia* spp.), sowthistle (*Sonchus oleraceous* L.), and wheat (*Triticum aestivum*). The most important summer and fall hosts were partridge pea (*Cassia fasciculata* Michaux [= *Chamaecrista fasciculata*]), wild lettuce (*Lactuca canadensis* L.), horseweed (*Erigeron canadensis* L.), and soybean. Other breeding hosts included spiderwort (*Tradescantia* spp.) and grasses. So, how does this explain the generally noneconomic status of this bug on soybean?

It is important to remember the number of host plants listed by Jones and Sullivan (1982) and that this insect is highly polyphagous (e.g., McPherson 1982). Jones and Sullivan (1982) suggested that because *Euschistus servus* has such a wide range of host plants, there is sufficient overlapping of uncultivated fruiting plants that large populations do not develop on soybean. McPherson et al. (1993) found in their study in Georgia conducted from mid-June to mid-October, 1987 to 1991, that population densities of *E. servus* on soybean remained relatively constant between years. This would suggest that because its range of host plants is so wide, the influence of the annual success of soybean production is less important for this stink bug's survival than it might be for other species (e.g., *Nezara viridula*).

5.3.5.2 Alfalfa

5.3.5.2.1 General Information
Euschistus servus will feed readily on the vegetative growth and developing seed of alfalfa (Russell 1952, Wene and Sheets 1964). In fact, Wene and Sheets (1964) stated that it preferred alfalfa to cotton. Overwintering adults prefer uncultivated, or weedy, host plants when they emerge from winter shelters (Russell 1952). They also may overwinter in cultivated land, particularly citrus groves, where winter or early spring vegetables are grown. These hosts bloom and produce seed from mid-January to late April or early May and serve as ovipositional sites for most of the overwintered females. The resulting nymphs reach maturity on these same plants, feeding mainly on the immature seeds. Nymphs begin to mature in late April and early May, the same time the weeds are drying up and are no longer suitable for feeding. Therefore, most of these adults move into fields of cultivated crops, particularly small grains, sugar beets, alfalfa, and cotton. Generally, they are present in fields of small grains and sugar beets from late April to late May but may be found in sugar beets until early July, the time depending on when these crops mature. Subsequently, they migrate to alfalfa fields that are producing seed. They remain there until the alfalfa plants mature or the crop is harvested, then migrate into later-maturing seed alfalfa or to fields of cotton. If a second crop of alfalfa is produced in August or September, the bugs often move in from an earlier alfalfa crop or from cotton. Alfalfa and cotton are important host plants from June to September and July to October, respectively. From early September to mid-October, the bugs may migrate to grain sorghum fields that are producing seed. Usually, the bugs are most common in alfalfa during the first half of July, shortly before the main seed crop reaches maturity. Following this, the bug population peaks on cotton during the last half of August and on grain sorghum during late September and early October (Russell 1952).

5.3.5.2.2 Damage
We have no specific information on damage to alfalfa by this stink bug.

5.3.5.3　Pecan

5.3.5.3.1　General Information

Euschistus servus feeds on pecan, attacking the nuts during their development (Adair 1927, 1932; Dutcher 1984; Dutcher and Todd 1983; Smith 1996a, b; Yonce and Mizell 1997) (Figure 5.1); apparently, pecan serves only as a feeding host (Adair 1932; Polles 1977, 1979; Stein 1985). Adults feed and/or reproduce on purple-flowered thistle (*Cirsium virginianum* L.), basketflower (*Centaurea americana* Nuttall), bean, cowpea, squash, tomato, peach, okra, soybean, pepper (= bell pepper?), cotton, corn, mesquite, and cocklebur (= common cocklebur?) (Adair 1932).

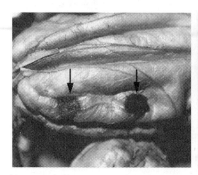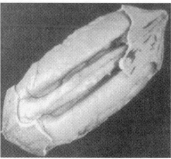

FIGURE 5.1　Damaged pecan (left) resulting from stink bug feeding, and undamaged pecan (right). (Courtesy of J.D. Dutcher, University of Georgia.)

Euschistus servus is bivoltine (e.g., Adair 1932) and overwinters as adults in or near the pecan orchards (Adair 1932, Dutcher and Todd 1983). Nymphs of the first generation become adults during June and July and those of the second generation from approximately late August to early October (Adair 1932). It is primarily adults of the second generation that enter orchards and feed on pecan (Yonce and Mizell 1997).

Adults have been found dead on thistle in early spring following cold temperatures that were preceded by several days of warm temperatures (Adair 1932).

5.3.5.3.2　Damage

This species causes black pit (Adair 1927, 1932; Dutcher and Todd 1983) and kernel spot (Dutcher and Todd 1983) to the nuts of pecan. For further information on these types of damage, see the discussion under *Nezara viridula*.

5.3.5.4　Sorghum

5.3.5.4.1　General Information

Although stink bugs are cosmopolitan, they are only occasional pests of sorghum, where their feeding on the developing heads produces smaller, lighter, and distorted seed (Young and Teetes 1977). One of these species is *Euschistus servus*.

5.3.5.4.2 Damage

Euschistus servus attacks the developing seeds of sorghum; specifically, it feeds on the kernels when they are in the milk (Hall and Teetes 1981), soft-dough (Hall and Teetes 1981, Wiseman and McMillian 1971), and hard-dough (Hall and Teetes 1981) stages of development. Wiseman and McMillian (1971) found up to 4 bugs (*E. servus* and *Nezara viridula*) on single heads and noted that adults and nymphs pierced several kernels. Hall and Teetes (1981) reported that concurrently with its collection on sorghum, *E. servus* also was found on *Cirsium texanum* Buckley, *Gaura villosa* Torrey, and cotton. Perhaps its low numbers on sorghum were because it also was attracted to alternate hosts nearby.

5.3.5.5 Corn

5.3.5.5.1 General Information

Although *Euschistus servus* can attack corn and cause damage, it generally is not considered a pest of this crop (e.g., Bergman 1999, Tonhasca and Stinner 1991). This probably is why so little information is available on the effects of its feeding. However, those studies that have been published suggest that this stink bug has the potential to be a pest. Injury is more likely when winters are mild, wheat or rye is used as a winter crop, and corn is planted without tillage (Bergman 1999).

Adults overwinter in vegetation in or near cultivated fields. Preferred overwintering habitats include wheat, alfalfa, or rye cover crops, or fall-seeded small grains. They emerge in spring and begin feeding and reproducing on alfalfa, wheat, and other host plants (Bergman 1999). They subsequently move to corn.

5.3.5.5.2 Damage

Adults and nymphs inject digestive enzymes and other compounds into the stems during feeding that can be phytotoxic or cause growth abnormalities (Bergman 1999). Feeding on young corn by *Euschistus servus* can affect growth and yield. For example, Townsend and Sedlacek (1986), from their greenhouse studies, found that a single adult feeding on seedling and stage 0.5 corn caused significant reduction in growth; feeding on stage 0.5 corn also produced tillering (shoot development from the base of a damaged plant). Apriyanto et al. (1989b) found that stage V2 corn attacked by single adults resulted in 52.5% of the plants forming tillers. Tillered plants had delayed silking and significantly reduced mean grain weight per ear. Sedlacek and Townsend (1988) reported that exposure of VE, V2, and V4 corn to bug densities of 1, 2, or 3 third instars, fourth and fifth instars combined, or adults resulted in termination of growth, reduced growth, or, with the exception of V4, mortality of plants. Nymphs and adults also caused tillering of VE, V2, and, with the exception of third instars, V4 corn. Apriyanto et al. (1989a) found the number of stylet sheaths (evidence of feeding or probing) increased with duration of exposure. Numbers of stylet sheaths per plant and percentage of damaged plants increased with increased insect density. There was a significant increase in sheaths as the observed plant damage increased (i.e., minor damage, growth deformity, tillering). Finally, Bergman (1999) stated that injury is most severe when planters are not adjusted properly, resulting in partially open seed slots that provide the bugs access to the underground stems and growing points of the young seedlings.

5.3.5.6 Peach

5.3.5.6.1   General Information

Peaches are attacked by several species of stink bugs, three of which are members of the genus *Euschistus* (i.e., *E. servus*, *E. tristigmus*, and *E. variolarius*) (Porter et al. 1928; Chandler 1943, 1950, 1955; Chandler and Flint 1939; Rings 1955; Woodside 1946a, b, 1947, 1950) (Figure 5.2). Chandler (1943, 1950) found that of the specimens of stink bugs jarred from the peach trees in the Illinois orchards with which he worked, *E. servus* made up approximately 50%.

FIGURE 5.2 *Euschistus* stink bug on peach. (Courtesy of H C Ellis, University of Georgia.)

Euschistus servus overwinters as adults that emerge in the spring and begin feeding and reproducing on various weeds; the bugs prefer ironweed, *Vernonia altissima* Nuttall (Rings 1955); white-top fleabane, *Erigeron annuus* (L.) (Woodside 1947, 1950); white campion, *Lychnis alba* Miller; common mullein, *Verbascum thapsus*; and horseweed, *Erigeron canadensis* L. (Rings 1955; Woodside 1947, 1950). In Virginia, high populations occur on white campion during May and June when this weed is in bloom (Woodside 1947). These populations are comprised of overwintered adults and their offspring. However, the favorite weed host of the first (summer) generation is white-top fleabane. This plant is attractive during June and July but as the seeds mature in July, the bugs migrate to more attractive hosts (Woodside 1947).

Horseweed, *Erigeron canadensis*, is the most important weed host of the second (overwintering) generation in central Virginia (Woodside 1947). This plant blooms from August to late September, and the bugs are most common on the plant at this time.

Peach orchards often are near woods, fence rows, ditches, and waste ground or, in other words, near the very areas that provide excellent overwintering sites and support a high diversity of weeds. Therefore, these orchards are vulnerable to attack from *Euschistus servus* (and other *Euschistus* spp.) when peaches are developing in the spring and, in fact, fruits on trees around the edges of orchards usually are attacked more severely by *E. servus* and other stink bugs early in the year (Woodside 1946b, 1949). Similarly, orchards that are not well managed and contain weeds among the trees are more subject to damage (Woodside 1949). Most bugs leave by the first of June (Woodside 1949).

5.3.5.6.2 Damage

Feeding by stink bugs on the fruits results in several types of damage (Rings 1955, 1957), two of which are catfacing and dimpling (Figures 5.3 and 5.4). The more severe damage is catfacing and is characterized by sunken areas surrounded by tissues that are more or less distorted (Porter et al. 1928); the surfaces are rough, corky, and lack pubescence (Woodside 1946b, 1950). Dimpling also is characterized by sunken areas, but the pubescence persists (Rings 1957; Woodside 1946b, 1950); it is considered an atypical type of catfacing (Rings 1957). These types of damage are related to the size of the peach when it is attacked. Fruits less than 3/8 in. diam. soon drop off after attack; those from 1/2 to 1 in. diam. develop typical catfacing, and those more than 1 inch in diam. develop a dimpled appearance (Woodside 1946b). Rings (1957) found that catfacing occurred when the developing fruits were attacked from bloom up to 49 days after bloom, and scarring (or dimpling) from 42 to 56 days after bloom.

5.3.5.7 Pear

5.3.5.7.1 Damage

Mundinger and Chapman (1932) reported several stink bugs feeding on pear, including *Euschistus servus*, in Athens, NY. They noted that the most damage occurred near woodlands or where an orchard bordered uncultivated land. The bugs apparently were attracted to the fruit as it ripened. Feeding on the pears resulted in shallow depressions beneath which were irregular white patches of pithy material.

5.3.5.8 Apple

5.3.5.8.1 General Information

Phillips and Howell (1980) studied the bionomics of *Euschistus servus* and *E. tristigmus* in apple orchards in Georgia, but, unfortunately, did not separate the data for the two species; thus, their results, which are given below, were based on combined data.

Both species are bivoltine and overwinter as adults. The bugs began to emerge in the spring and were collected from the apple trees from early to mid-April and

FIGURE 5.3 Catfacing on peach resulting from stink bug feeding. (Courtesy of the Clemson University Extension Service slide exchange program.)

from early June to mid-August. Orchard ground cover was green by the end of April, and the bugs apparently moved to this vegetation from the trees. The first adult was collected from ground cover on May 2 ; the highest density of adults in this vegetation (June) coincided with the highest density of adults on the trees.

Nymphs of the first generation were swept from ground cover from the second week of May to the end of June, peaking between May 16 and 31. The ground cover had dried up in June, but the weeds in adjoining fields remained green. Adults and nymphs were swept from these fields in June and July. Both months were hot and

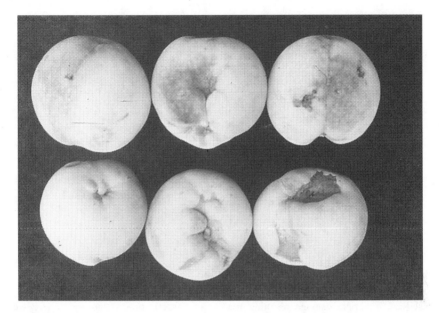

FIGURE 5.4 Various stages of catfacing and dimpling on peach resulting from stink bug feeding. (Courtesy of J.R. Meyer, North Carolina State University.)

dry but were followed by several days of rain at the end of July and the beginning of August. As a result, the ground cover began to recover. Nymphs of the second generation were found on the renewed cover, and these nymphs reached adults during September.

5.3.5.8.2 Damage
Phillips and Howell (1980) found there was a correlation between the number of adults and nymphs jarred from the trees and the percent of damaged fruit. Damage was characterized by dried-out, light-brown corky areas (Figure 5.5). They suggested that the amount of damage in an orchard might be related directly to the composition and density of the orchard ground cover and the type of vegetation adjacent to the orchard.

5.3.5.9 Cotton

5.3.5.9.1 General Information
One of the earliest reports of *Euschistus servus* attacking cotton was that of Morrill (1910) who referred to this insect as the "Brown cotton-bug." He stated that, in Texas, weeds along roadsides and fences undoubtedly provided favorable breeding places for the bugs in early summer but that after appearance of the cotton bolls, comparatively few bugs could be found outside the cotton field. Clancy (1946) stated that this bug was the most serious insect pest in Yuma Valley, AZ, and that its infestations were characterized by rapid shifts from other host plants to cotton as the bolls were forming. Wene and Sheets (1964) also considered it an important pest of cotton in Arizona, particularly in the central part of the state. Turnipseed et al.

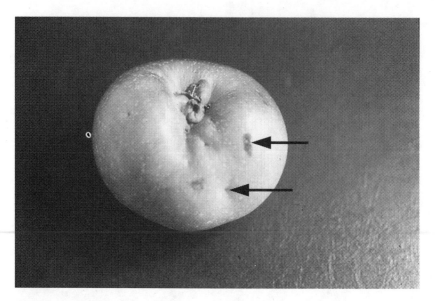

FIGURE 5.5 Apple damage resulting from stink bug feeding. (Courtesy of H C Ellis, University of Georgia.)

(1995) noted that it and other stink bugs were major secondary pests of cotton in South Carolina; Bundy and McPherson (2000b) reported similar findings in Georgia.

Barbour et al. (1988) found that in North Carolina, *Euschistus servus* adults initially occurred in cotton in approximately mid-June, increased in numbers by mid-July, and peaked in early August. Nymphs were found soon after August 1.

5.3.5.9.2 Damage

Barbour et al. (1988), from greenhouse tests in which they used *Euschistus servus* adults caged individually in small net cages on bolls 1- to 2-weeks old, found a reduction in the number of harvestable locks. Roach (1988), using field cages containing individual cotton plants and 0, 1, or 2 adults, found the bugs appeared to have reduced the number of harvestable locks remaining on the plants, but the results had been obscured by weather factors such as drought.

The bugs feed on the bolls, inserting their beaks into the developing seeds and surrounding tissues. Feeding on young bolls causes them to become soft and yellow and to shed. Feeding on older bolls causes them to form rough, warty, cellular growths on the carpels. Fungi may enter through feeding punctures, causing the locks to decay (Wene and Sheets 1964).

Turnipseed et al. (1995) reported that *Euschistus servus*, *Nezara viridula*, and *Acrosternum hilare* were the major secondary pests observed in untreated Bt cotton during their study of Bt cotton in South Carolina. They found 7% damaged bolls in non-Bt cotton treated twice with Karate® insecticide compared with 27% damaged bolls in untreated Bt cotton.

Euschistus servus appears to be increasing in importance as a pest of cotton. This apparently is related directly, at least in part, to current control efforts of the boll weevil (see discussion under *Acrosternum hilare*).

5.3.5.10 Other Crops

5.3.5.10.1 Damage

Euschistus servus has been reported from tomato (Callahan et al. 1960), sugar beets (Hills and McKinney 1946), and tobacco (Jewett 1959, Rabb et al. 1959, Reich 1991, Woodside 1947). Callahan et al. (1960) reported that both nymphs and adults feed on green fruits of tomato, causing the fruits to have a bitter taste and pithy texture. Hills and McKinney (1946), based on field-cage studies, have found that this bug is capable of reducing seed germination in sugar beets and the number of sprouts per viable seed ball.

5.4 *EUSCHISTUS VARIOLARIUS* (PALISOT DE BEAUVOIS)

5.4.1 INTRODUCTION

Euschistus variolarius, the one spotted stink bug, occurs from Quebec, Ontario, and New England south to Florida and west to British Columbia, Oregon, and Utah (Froeschner 1988).

This stink bug is highly polyphagous, feeding on numerous cultivated and uncultivated plants including soybean, to which it is capable of transmitting yeast-spot disease (caused by *Nematospora coryli*) under caged conditions (McPherson 1982). Although generally regarded as phytophagous, it also has been reported as predaceous on other insects including species of Homoptera, Coleoptera, and Lepidoptera (McPherson 1982). However, Esselbaugh (1948) suggested these records might be based on misidentifications of this stink bug because the adults superficially resemble those of the predaceous stink bug *Podisus maculiventris*.

5.4.2 LIFE HISTORY

Euschistus variolarius overwinters as adults (McPherson 1982, Munyaneza and McPherson 1994) and is listed as both univoltine (Munyaneza and McPherson 1994; Mundinger 1940; Parish 1934; Rings and Brooks 1958; Woodside 1946a, 1950) and bivoltine (Esselbaugh 1948, Wilks 1964?).

5.4.2.1 Mating Behavior

Precopulatory and copulatory behavior has been studied in this species (Drickamer and McPherson 1992). Touching is initiated by the male, who usually begins by antennating the female's head, with or without head contact. He then moves along her side, antennating and making head contact, until reaching her abdomen. Then, after further touching her abdomen, he turns and, without maintaining contact, attempts copulation in the end-to-end position.

5.4.3 LABORATORY INVESTIGATIONS

Euschistus variolarius has been reared under confined conditions (Esselbaugh 1948, Foot and Strobell 1914, Mundinger 1940, Munyaneza and McPherson 1994, Olsen 1912, Parish 1934, Rings and Brooks 1958, Sailer 1952, Scheel et al. 1957, Woodside 1946a).

5.4.4 DESCRIPTIONS OF IMMATURE STAGES

The eggs (Esselbaugh 1946, Mundinger 1940, Munyaneza and McPherson 1994, Parish 1934) and first to fifth instars (DeCoursey and Esselbaugh 1962, Mundinger 1940, Munyaneza and McPherson 1994, Parish 1934) have been described. Munyaneza and McPherson's study also included comparisons of the egg and first to fifth instars of *E. variolarius* with the corresponding stages of *E. servus*, which they closely resemble.

5.4.5 ECONOMIC IMPORTANCE

5.4.5.1 Soybean

5.4.5.1.1 Damage

Bickenstaff and Huggans (1962), using a mixture of *Euschistus variolarius* and *E. servus* in field cages containing soybean plants at various stages of development beginning at full bloom, found the plants showed progressively less damage as they matured from full bloom or just past bloom to leaf drop.

5.4.5.2 Corn

5.4.5.2.1 General Information

Although *Euschistus variolarius* can attack corn and cause damage, it generally is not considered a pest of this crop (Bergman 1999). This probably is why so little information is available on the effects of its feeding. However, those studies that have been published suggest that this stink bug has the potential to be a pest. Injury is more likely when winters are mild, wheat or rye is used as a winter crop, and corn is planted without tillage (Bergman 1999).

Adults overwinter in vegetation in or near cultivated fields. Preferred overwintering habitats include wheat, alfalfa, or rye cover crops, or fall-seeded small grains. They emerge in spring and begin feeding and reproducing on alfalfa, wheat, and other host plants (Bergman 1999). They subsequently move to corn.

5.4.5.2.2 Damage

Adults and nymphs inject digestive enzymes and other compounds into the stems during feeding that can be phytotoxic or cause growth abnormalities (Bergman 1999). Feeding on young corn by *Euschistus variolarius* can affect growth and yield. For example, Townsend and Sedlacek (1986), from their greenhouse studies, found that a single adult feeding on seedling and stage 0.5 corn caused significant reduction in growth; feeding on stage 0.5 corn also produced tillering (shoot development from the base of a damaged plant) but was more severe in plants exposed to *Euschistus servus*. Apriyanto et al. (1989b) reported that stage V2 corn attacked by single adults resulted in 38.8% of the plants forming tillers. Tillered plants had delayed silking and significantly reduced mean grain weight per ear. Sedlacek and Townsend (1988) reported that exposure of VE, V2, and V4 corn to bug densities of 1, 2, or 3 third instars, fourth and fifth instars combined, or adults resulted in termination of growth, reduced growth, or, with the exception of V4, mortality of plants. Nymphs and adults also caused tillering of VE, V2, and, with the exception of third instars, V4 corn. Apriyanto et al. (1989a) found the number of stylet sheaths (evidence of feeding or probing) increased with duration of exposure. Annan and Bergman (1988) reported that exposure of corn at the VE, V2, and V4 stages to single fourth instars or adults resulted in significant yield reductions, growth deformities, and, occasionally, death. However, those plants infested at the V6 stage under greenhouse conditions rapidly outgrew the effects of stink bug feeding. Finally, Bergman (1999) stated that injury is most severe when planters are not adjusted properly, resulting in partially open seed slots that provide the bugs access to the underground stems and growing points of the young seedlings.

5.4.5.3   Peach

5.4.5.3.1   Damage
See *Euschistus servus* for description of damage.

5.4.5.4   Pear

5.4.5.4.1   Damage
Mundinger and Chapman (1932) reported several stink bugs feeding on pear, including *Euschistus variolarius*, in Athens, NY. They noted that most damage occurred near woodlands or where an orchard bordered uncultivated land. The bugs apparently were attracted to the fruit as it ripened. Feeding on the pears resulted in shallow depressions beneath which were irregular white patches of pithy material.

Several years later, Wilks (1964) reported on the damage to pear called "cottony spot" in British Columbia. He described the damage as small white cottony pockets, approximately 1/4 in. diam., that occurred just beneath the skin and extended approximately 1/8 in. into the flesh. He correlated the damage to the life cycle of *Euschistus variolarius* and proved that it was the result of feeding by this bug. He observed two population peaks, one in late May and early June, which occurred mainly in the orchard cover crop; and a second peak in early August, which occurred in both the cover crop and in pear trees. He noted that most injury appeared to occur in August during the second population peak. Also, during the second peak, he observed the bugs feeding frequently under the leaves near the stem end of the fruit, the area where most cottony spot injury occurs. He proved the bugs were responsible for cottony spot, as follows: he collected nymphs and adults in early to mid-August from the cover crop; then he caged branches with fruit, added bugs to some of these cages, the other cages serving as controls, and found after several weeks that only those pears with bugs developed cottony spot.

Although Wilks (1964) did not state that this stink bug was bivoltine, his results (two peaks) suggest that it was, depending on whether the first peak represented overwintering bugs or their offspring.

5.5   *EUSCHISTUS TRISTIGMUS* (SAY)

5.5.1   Introduction

Euschistus tristigmus, the dusky stink bug, occurs across much of North America and is comprised of two subspecies, *E. t. luridus* Dallas, which occurs primarily north, and *E. t. tristigmus*, which occurs primarily south, of latitude 41°N (McPherson 1982). The species is highly polyphagous, attacking both uncultivated and cultivated plants (McPherson 1982). Some of the more important crops attacked include soybean (Jones and Sullivan 1979; McPherson et al. 1979a, b), to which it is capable of transmitting yeast-spot disease (caused by *Nematospora coryli*) under caged conditions (Daugherty 1967), peach (Chandler 1943, 1950, 1955; Chandler and Flint 1939; Furth 1974; Porter et al. 1928; Rings 1955, 1957; Woodside 1946a, b, 1949, 1950), apple (Phillips and Howell 1980), snap bean, lima bean (Woodside 1947), pear (Mundinger and Chapman 1932), and tomato (Mundinger 1940, Woodside 1947). Often, it is associated with *Euschistus servus* and/or *E. variolarius* but seems to be of secondary importance when compared with the other two. Also, it will feed on larvae of the eyespotted bud moth, *Spilonota ocellana* (Denis and Schiffermüller), under caged conditions (Stultz 1955).

5.5.2   Life History

Euschistus t. tristigmus (and, perhaps, *E. t. luridus*) overwinters as adults (McPherson 1975, 1982; McPherson and Mohlenbrock 1976); it appears to prefer deciduous woods and their borders as overwintering sites (Jones and Sullivan 1981). It has been reported as bivoltine in Illinois (McPherson 1975, McPherson and Mohlenbrock 1976), Virginia (Woodside 1946a, 1947, 1950), and Georgia (Phillips and Howell 1980), but has, perhaps, only two partial generations around Urbana, IL (Esselbaugh 1948).

5.5.2.1   Mating Behavior

Precopulatory and copulatory behavior has been studied in this species (Clair and McPherson 1980, Drickamer and McPherson 1992). Touching behavior is initiated by the male, who usually begins by antennating the female's head, thorax, or abdomen, moving head-to-rear or vice versa. He eventually concentrates his antennating along the ventral side of her abdomen and usually lifts her posterior end with his head. If receptive, the female keeps her abdomen elevated. The male then rotates 180°, elevates his abdomen, and attempts copulation. Occasionally, the male will maintain contact with the female during rotation by sliding his abdomen along her side until achieving end-to-end contact. During copulation, the male often rests his hind tarsi on her hind legs.

5.5.3 Laboratory Investigations

Euschistus tristigmus has been reared under confined conditions from egg to adult (Clair and McPherson 1980, Esselbaugh 1948, McPherson 1971, Sailer 1952, Woodside 1946a).

5.5.4 Descriptions of Immature Stages

The eggs (Esselbaugh 1946, Javahery 1994) and first to fifth instars (DeCoursey and Esselbaugh 1962) have been described.

5.5.5 Economic Importance

5.5.5.1 Soybean

5.5.5.1.1 Damage
Euschistus tristigmus often has been reported from soybean but is considered to be of minor importance (e.g., Jones and Sullivan 1983). However, McPherson et al. (1979a) reported that under field-cage conditions, it caused as much damage to soybeans as third instar *Nezara viridula*.

5.5.5.2 Peach

5.5.5.2.1 General Information
Euschistus tristigmus overwinters as adults that emerge in the spring and begin feeding and reproducing on various weeds; the bugs prefer ironweed, *Vernonia altissima* (Rings 1955); white-top fleabane, *Erigeron annuus* (Woodside 1947, 1950); white campion, *Lychnis alba* Miller; mullein, *Verbascum thapsus*; and horseweed, *Erigeron canadensis* (Rings 1955; Woodside 1947, 1950). In Virginia, high populations occur on white campion during May and June when this weed is in bloom (Woodside 1947). These populations are comprised of overwintered adults and their offspring. However, the favorite weed host of the first (summer) generation is white-top fleabane. This plant is attractive during June and July, but as the seeds mature in July, the bugs migrate to more attractive hosts (Woodside 1947).

Horseweed, *Erigeron canadensis*, is the most important weed host of the second (overwintering) generation in central Virginia (Woodside 1947). This plant blooms from August to late September, and the bugs are most common on the plant at this time.

Peach orchards are often near woods, fence rows, ditches, and waste ground or, in other words, near the very areas that provide excellent overwintering sites and support a high diversity of weeds. Therefore, these orchards are vulnerable to attack from *Euschistus tristigmus* (and other *Euschistus* spp.) when peaches are developing in the spring; in fact, fruits on trees around the edges of orchards usually are attacked more severely by *E. tristigmus* and other stink bugs early in the year (Woodside 1946b, 1949). Similarly, orchards that are not well managed and contain weeds among the trees are more subject to damage (Woodside 1949). Most bugs leave by the first of June (Woodside 1949).

5.5.5.2.2   Damage

See *Euschistus servus* for description of damage.

5.5.5.3   Pear

5.5.5.3.1   Damage

Mundinger and Chapman (1932) reported several stink bugs feeding on pear, including *Euschistus tristigmus*, in Athens, NY. They noted that most damage occurred in the vicinity of woodlands or where an orchard bordered uncultivated land. The bugs apparently were attracted to the fruit as it ripened. Feeding on the pears resulted in shallow depressions beneath which were irregular white patches of pithy material.

5.5.5.4   Apple

5.5.5.4.1   Damage

See Phillips and Howell (1980) under *Euschistus servus*.

5.5.5.5   Pecan

5.5.5.5.1   Damage

Euschistus tristigmus has been listed as one of a complex of stink bugs attacking pecan, the others being *Euschistus servus*, *Acrosternum hilare*, and *Nezara viridula* (Mizell et al. 1997; Smith 1996a, b; Yonce and Mizell 1997). We have no information on damage specifically attributable to this stink bug.

5.6   *EUSCHISTUS CONSPERSUS* UHLER

5.6.1   Introduction

Euschistus conspersus has been recorded from California, Washington, British Columbia (Froeschner 1988), Oregon, Nevada, and Idaho (Borden et al. 1952).

This bug attacks several cultivated crops including cotton (Toscano and Stern 1976a, b), alfalfa (Scott and Madsen 1950, Toscano and Stern 1976b), sorghum, sugar beets (Toscano and Stern 1976b), tomatoes (Borden et al. 1952, Hoffmann et al. 1987b, Zalom and Zalom 1992, Zalom et al. 1997a, b), peaches (Borden et al. 1952, Perry 1979), pear (Barnett et al. 1976, Borden and Madsen 1953, Borden et al. 1952, Perry 1979, Scott and Madsen 1950, Swan and Papp 1972), apricots, figs, plums, apple (Borden et al. 1952), and berries (Alcock 1971, Essig 1926, Perry 1979, Scott and Madsen 1950). Uncultivated host plants include dock (Barnett et al. 1976), mallow or cheeseweed (*Malva parviflora* L.), radish (*Raphanus sativus* L.), black mustard [*Brassica nigra* (L.)], and common mustard [*Brassica kaber* (DC)] (Ehler 2000).

5.6.2   Life History

Euschistus conspersus has been reported as univoltine (Alcock 1971), but most authors have considered it bivoltine (Barnett et al. 1976, Borden and Madsen 1953,

Borden et al. 1952, Ehler 2000, Toscano and Stern 1980). Adults overwinter in weeds, trash, dead grass, and leaf litter and emerge in the spring (Alcock 1971, Barnett et al. 1976, Borden and Madsen 1953, Borden et al. 1952, Ehler 2000, Scott and Madsen 1950, Toscano and Stern 1980). They feed and reproduce on broadleaf plants (e.g., dock [*Rumex*], yellow mustard, black mustard, radish) (Barnett et al. 1976, Borden and Madsen 1953, Borden et al. 1952, Ehler 2000), and their offspring mature during June and July. These new adults (summer generation) reproduce during July and August, giving rise to the second (overwintering) generation (Barnett et al. 1976, Borden et al. 1952, Ehler 2000, Toscano and Stern 1980).

5.6.2.1 Mating Behavior

Alcock (1971) studied the precopulatory and copulatory behavior of this stink bug. He reported that during precopulation, the male antennates the dorsal side of the female and attempts to lift her abdomen with his head. Subsequently, he turns and presses his aedeagus, which may be completely extruded and partially inverted, against the sides, venter, or tip of her abdomen. An unresponsive female will not lift her abdomen, or will move her abdomen slowly from side to side, rock back and forth as the male presses against her, or push the male away with a hind leg. Males are persistent, however, and will repeat the courting process over and over. If the female is responsive, she will slowly lift the tip of her abdomen. The male will continue to court as above but concentrate particularly on lifting her abdomen and touching her with his aedeagus. In addition, he now will antennate the venter of her abdomen while standing behind her. When she has lifted the tip of her abdomen to approximately 30°, he will turn 180° and, with his aedeagus completely extruded and inverted, back toward her with his abdomen raised and insert his aedeagus.

5.6.3 LABORATORY INVESTIGATIONS

Euschistus conspersus has been reared in the laboratory from egg to adult (Cullen and Zalom 2000, Ehler 2000, Hunter and Leigh 1965, Toscano and Stern 1976c), and the effects of various temperatures on the incubation period, nymphal stadia, preovipositional and ovipositional periods, and fecundity have been investigated (Toscano and Stern 1976c). In addition, Toscano and Stern (1980) have found that it is possible to predict seasonal patterns in populations by changes in the female's reproductive system.

5.6.4 ECONOMIC IMPORTANCE

5.6.4.1 Tomato

5.6.4.1.1 General Information

Euschistus conspersus frequently attacks tomatoes (Figures 5.6 to 5.8) and, in fact, is the species in California observed damaging processing tomatoes most frequently (Hoffmann et al. 1987b).

Adults overwinter in and around orchards under the leaves of legumes, blackberries, and weeds such as Russian thistle and mallow. They emerge in early spring

FIGURE 5.6 *Euschistus* stink bug feeding on green tomato. (Courtesy of D.G. Riley, University of Georgia.)

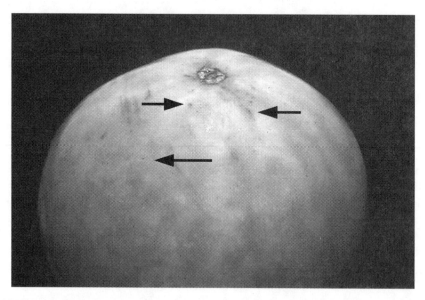

FIGURE 5.7 Pinprick damage on green tomato resulting from stink bug feeding. (From Anonymous. 1998. Integrated Pest Management for Tomatoes (4th ed.) IPM Education and Publications, University of California, Davis. With permission.)

FIGURE 5.8 Irregular blotches on ripe tomato resulting from stink bug feeding. (Anonymous. 1998. Integrated Pest Management for Tomatoes (4th ed.) IPM Education and Publications, University of California, Davis. With permission.)

and begin feeding and reproducing on weeds and other host crops, but often are found along the edges of sloughs and creeks where blackberries occur (Anonymous 1998). Other wild hosts include black mustard [*Brassica nigra* (L.)] and radish (*Raphanus sativus* L.) (Ehler 2000).

This species is bivoltine. The most serious damage to tomatoes results from the second generation, which migrates into the tomatoes from surrounding areas, feeds on the fruit, and overwinters (Anonymous 1998, Ehler 2000).

Hoffmann et al. (1987b) studied movement of these bugs on caged tomato plants in the field, noting their distribution at approximately 7:00 and 10:00 A.M. and 3:00 and 7:30 P.M. They found no significant shift in the vertical distribution of the bugs during the day; throughout the day, more than 50% of the bugs was on the plant (including those on the cage walls), approximately 30% was on the soil surface, and the remainder was in the soil. They also observed the bugs' distribution at sunrise and sunset; they found that all moved off the soil when it was wet (e.g., as within a few days following irrigation), whereas 44% was found on or in the soil when it was dry. Zalom et al. (1997b) studied field spatial distribution and developed a sampling procedure for estimating population density that took spatial distribution into account.

5.6.4.1.2 Damage
Damage to fruit often is limited to the sides of fields because the stink bugs migrate from surrounding host plants; however, in years with heavy spring rain and late

weed growth, the bug populations may be high and the damage in the fields more widespread (Anonymous 1998). Michelbacher et al. (1952) reported severe damage to a field of tomatoes in Sacramento, CA. The tomato field was examined on August 26; the half that was located next to a fescue seed field, which had been harvested approximately July 1, had a high proportion of injured green and ripe fruits (presumably, the other half of the tomato field had fewer injured fruits). Although not stated, these observations suggest that the bugs had moved from the fescue to the tomato. All stages of the bug were present, indicating that tomato was a suitable host for development. Furthermore, green fruit apparently was preferred.

Feeding by these bugs results in destruction of cells just beneath the surface (Michelbacher et al. 1952). Below this point, the tissue appears normal. When green tomatoes are attacked severely, the fruit appears mottled and is relative spongy because of destruction of the subsurface tissue. The destroyed tissue is nearly white and cottony in appearance. The mottled appearance persists as the fruit ripens; the injured area remains soft and spongy; and the injured tissue remains white and cottony (Figure 5.9). If the damage on the green fruit is not severe, some recovery is possible (Michelbacher et al. 1952). Damage apparently results from the injection of a digestive enzyme during feeding and the withdrawal of cellular fluids; also, yeasts and bacteria associated with the bugs' mouthparts may cause decay when introduced into the fruits (Anonymous 1998).

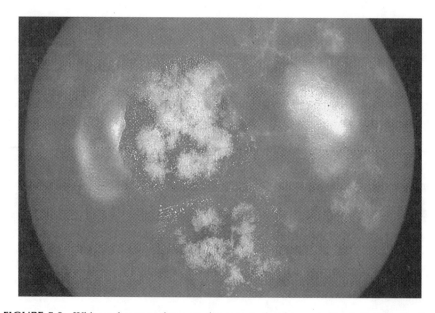

FIGURE 5.9 White and spongy tissue on ripe tomato that forms at site of stink bug feeding punctures. (Anonymous. 1998. Integrated Pest Management for Tomatoes (4th ed.) IPM Education and Publications, University of California, Davis. With permission.)

5.6.4.2 Pear

5.6.4.2.1 *General Information*

Euschistus conspersus feeds on pear and can cause economic damage (Barnett et al. 1976, Borden and Madsen 1953, Borden et al. 1952, Scott and Madsen 1950, Swan and Papp 1972). However, it also will attack peaches, plums, apples, apricots, and figs if these are interplanted with pears in infested orchards (Borden et al. 1952).

Adults overwinter in orchards from November to March under weeds or trash (Borden et al. 1952). In orchards with cover crops, they prefer dead grass and weeds under the trees. Overwintered adults emerge in early spring and begin feeding on green broadleaf plants in the cover crop, especially dock (*Rumex*) and yellow mustard. They lay their eggs shortly thereafter on the undersides of leaves of these same cover crop plants including dock, mustard, horehound, plantain (= broad-leaved plantain?), mallow, and other weeds. Concurrently, they migrate to blackberry (*Rubus*), mullein, thistle, and other green plants outside the orchard where they also lay eggs. Eggs also are laid on pear leaves but proportionally fewer than on green plants inside and outside the orchards.

As noted earlier, this species is bivoltine. Borden et al. (1952) reported that the incubation period of the eggs of the first (overwintered) generation ranged from 7 to 30 days and the nymphal period from 67 to 79 days. These nymphs fed on succulent broadleaf plants during the spring but could not develop on pear leaves, even if they hatched from eggs laid on pear foliage. During this time, the overwintered adults fed to some extent on the young pear fruit but, apparently, did little damage. Their adult offspring (summer generation) reached maturity by mid-June. Adults of this same generation developing outside the orchards migrated to green areas in the orchards as their host plants began to dry up in midsummer. These summer adults fed on the pears as well as on the cover crop, and it was during this time that the nearly mature fruit sustained the most damage from the bugs. The bugs reproduced, giving rise to the second (overwintering) generation. They oviposited on leaves of the fruit trees and cover crops and even on ground trash. Undoubtedly because of higher temperatures, the incubation period of these eggs ranged from 5 to 9 days and the nymphal period from 40 to 65 days. The resulting nymphs fed almost entirely on succulent plants. When they reached adults, the pears had been harvested and, thus, there was no damage from this generation.

5.6.4.2.2 *Damage*

Adults of the first generation feed at the stem end of the nearly mature pear, gradually extending the feeding area toward the middle of the fruit (Borden et al. 1952). The feeding punctures cause white, corky areas under the surface that may reach 1/2 in. into the flesh. The damaged areas turn brown when exposed to air. Heavy feeding may cause dimpling or distortion of the mature fruit (Borden et al. 1952). During feeding, the adults often deposit small amounts of dark excrement on the fruits and leaves (Barnett et al. 1976).

Borden et al. (1952) noted that in some orchards that were heavily infested, the two outside rows of pears sustained up to 90% damaged fruit, indicating the bugs

were migrating into the orchard from surrounding weedy areas. In comparison, fruit
that was farther inside the orchard showed damage of 10 to 50%.

5.6.4.3   Apple and Apricot

5.6.4.3.1   Damage

Euschistus conspersus attacks these crops, causing feeding damage in apple characterized by dark green spots under the skin and in apricot by flesh that turns white
and is visible through the thin skin of mature fruit (Borden et al. 1952).

5.6.4.4   Cotton (Alfalfa, Sorghum, Sugar Beet)

5.6.4.4.1   General Information

Euschistus conspersus attacks cotton and is capable of reducing yield and quality.
Apparently, however, cotton is not a preferred host, this based on the rarity of
nymphs in this crop compared with alfalfa (Toscano and Stern 1976b). Toscano
and Stern (1976b) conducted a study in the San Joaquin Valley, CA, in an area
comprised of four fields: cotton, sorghum, and sugar beet, each field bordering the
same alfalfa seed field. The first measurable number of bugs was found in the
alfalfa about mid-July when approximately 90% of the plants were in bloom.
Between then and seed set (mid-August), the bug's population (adults and nymphs)
increased rapidly. Just prior to harvest, when the alfalfa seed was mature, the
number of bugs dropped and the bugs began appearing in the cotton fields. As the
alfalfa seed was being harvested, the numbers of adults in cotton rose to their
highest levels (August 24), with most bugs collected along the edge of the cotton
field adjacent to the alfalfa field.

Population increase in the sorghum field was similar to that in the cotton field,
in that it was most apparent after alfalfa was no longer suitable as a food source.
However, population numbers in sugar beets did not show the same pattern. The
bugs were found in sugar beets between late June and late July; those found early
in the summer may have been remnants of an overwintering population (Toscano
and Stern 1976b).

It appears, based on the earlier discussion of this insect's general life cycle, that
the bugs entering the alfalfa field in mid-July were the offspring (summer generation)
of the overwintering adults, and the subsequent rapid increase in numbers in this
crop resulted from their offspring (second [= overwintering] generation). This is
supported by the rarity of nymphs in cotton as compared with alfalfa when the bugs
moved into cotton in August.

5.6.4.4.2   Damage

Toscano and Stern (1976a), using field cages and bug densities of 0 (check), 2, 4,
and 8 adults per cotton plant (approximately 0.9 to 1.1 m tall, bolls approximately
1.9 to 3.8 cm) per cage found that damage from the highest density (8) was most
severe and resulted in reduction of seed and lint weights, and in germination.

Confining reproductively active adult males and females in larger cages, each cage containing an average of 98 cotton plants and either 0, 25, 50, or 100 bugs, resulted in significant damage at the highest density (100), with a decrease in the number of harvestable bolls. Specifically, damage at this density caused a significant decrease in mean seed weight; it also caused a decrease in the percent embryo, protein in the cotton meal, and oil content, and an increase in the percent hull of the cotton seed.

6 *Acrosternum hilare* (Say)

CONTENTS

6.1 INTRODUCTION

Acrosternum hilare, the green stink bug, ranges from Quebec and New England west through southern Canada and the northern United States to the Pacific Coast and south and southwest to Florida, Texas, Arizona, Utah, and California (McPherson 1982). It can be confused with *Nezara viridula* but can be distinguished most easily by the long ostiolar canal, which extends well beyond the middle of its supporting plate; that of *N. viridula* is much shorter, not reaching the middle of its supporting plate.

    This insect is highly polyphagous but shows a distinct preference for woody shrubs and trees (Jones and Sullivan 1982, McPherson 1982). *Nezara viridula*, in contrast, prefers herbaceous annuals (Jones and Sullivan 1982), although it will attack trees (e.g., pecan, macadamia).

6.2 LIFE HISTORY

Acrosternum hilare, as with most stink bugs, overwinters as adults (McPherson 1982, McPherson and Tecic 1997). It seems to prefer leaf litter of deciduous woods (Javahery 1990; Jones and Sullivan 1981; Sorenson and Anthon 1936; Underhill 1934; Whitmarsh 1914, 1917), although other sites have been mentioned (e.g., beneath bark of fallen trees and behind corn stalk sheaths in Arkansas [Miner 1966a]). It has been reported as univoltine by most authors (e.g., Esselbaugh 1948; Javahery 1990; Schoene and Underhill 1933; Sorenson and Anthon 1936; Underhill 1934; Whitmarsh 1914, 1917) although Miner (1966a, b), Jones and Sullivan (1982), and Wilde (1969) considered it bivoltine in Arkansas, South Carolina, and Kansas, respectively. This difference in opinion (i.e., uni- or bivoltine) probably reflects differences in environmental conditions between the geographic areas in which these studies were conducted because, as Sailer (1953) suggested, the species might be bivoltine under favorable conditions such as those encountered in the Gulf States. Work in southern Illinois supports this contention. Southern Illinois lies within a broad south–north transition zone between the geographic locations of these earlier studies. McPherson and Mohlenbrock (1976) stated that this stink bug was apparently bivoltine in southern Illinois but could not be more definite because their data were inconclusive. Recently, however, McPherson and Tecic (1997), based on several years of additional work, have stated that, in fact, it is bivoltine in southern Illinois.

A sequence of host plants with overlapping periods of seed and fruit production is necessary for *Acrosternum hilare* to develop large populations during its annual life cycle. Schoene and Underhill (1933) and Underhill (1934) listed several host plants in Virginia and diagrammed when the seed pods and fruits were in suitable condition for feeding. Underhill (1934) stated that elderberry, black locust, honey locust, and mimosa were an excellent combination for population buildup. Schoene and Underhill (1933) mentioned that severe damage to farm or garden crops occurred when host succession was incomplete or broken and the cultivated crop was nearby and in suitable condition as food. Miner (1966a, b), in Arkansas, reported that a first generation occurred on wild hosts, especially dogwood (*Cornus drummondi* Myer), and that a second generation occurred on soybean; later infestations were higher on beans if the beans were nearer wild hosts. Jones and Sullivan (1982), in South Carolina, also presented a diagram showing the plants that *A. hilare* attacked and the seasonal sequence of the reproductive periods for those hosts. Some of the important wild host plants included photinia, American holly, Chinese privet, and trumpet-creeper. They stated that a first generation occurred on black cherry and elderberry and a second on soybean, and that the first two hosts allowed large populations to develop before the insects entered soybean fields. Woodside (1950), in Virginia, noted that adults, upon leaving overwintering sites, could go to peach immediately and deposit their eggs. The resulting nymphs were responsible for dimpling of the peaches, and the damage to individual fruits was severe. However, if many wild hosts were present near the orchard, the bugs would migrate from these hosts to the peach in July. Under these circumstances, severe damage usually resulted but could be limited to only a few trees. Finally, Sorenson and Anthon (1936), in

Utah, reported that this stink bug infested cherries, apricots, peaches, grapes, and pears in orchards in the order in which the fruits began to ripen.

6.2.1 MATING BEHAVIOR

Precopulatory and copulatory behavior of *Acrosternum hilare* has been studied (Javahery 1990, Whitmarsh 1917). Javahery (1990) reported that the male begins courtship by antennating the female's antennae, head, or lateral margins. The female, if receptive, remains on the substrate and lowers her head. The male then moves posteriorly along her side while continuing to antennate her until he reaches her posterior end. He then attempts to raise her posterior end with his head. If she is receptive, she raises the end of her abdomen approximately 40° and slightly opens her genital segments. The male then moves away approximately 1 cm, turns 180°, backs toward her with his pygophore twisted and aedeagus extended, raises his abdomen, and attempts copulation.

Capone (1995) found that females preferred to mate with larger males if the effect of direct physical male–male competition was removed, and males preferred to mate with larger females. She also found a positive relationship between female body size and fecundity. Thus, males have a direct reproductive gain by choosing larger females, but the advantage to females is not clear.

Lockwood and Story (1985c) catalogued the diurnal behavior of adult *Acrosternum hilare* in a field of senescing soybeans. They described 20 behaviors and quantified the effects of sex and time of day on these behaviors.

6.2.2 SCENT GLAND SECRETIONS

Aldrich (1995a, 1996) and Aldrich et al. (1989) reported that gas chromatograms of airborne-trapped volatiles for this stink bug exhibit a male-specific pattern qualitatively similar to that of *Nezara viridula* from the eastern United States but with the abundances of the (Z)-α-bisabolene epoxide isomers reversed; they discussed the components of this pheromone blend in further detail. ·

Recently, Aldrich et al. (1995) analyzed the dorsal abdominal gland secretions of several stink bug species including *Acrosternum hilare*. They identified four volatiles in *A. hilare* including hexanal, (E)-2-hexenal, hexanoic acid, and (E)-2-hexenoic acid. At the concentrations tested, they concluded that these secretions might be epideictic, promoting spacing in the natural environment.

6.3 LABORATORY INVESTIGATIONS

This species has been reared under confined conditions by several individuals (Esselbaugh 1948; Menusan 1943; Miner 1966a, b; Russell 1952; Sailer 1952, 1953; Sorenson and Anthon 1936; Underhill 1934; Whitmarsh 1914, 1917; Wilde 1968, 1969), the most recent in which stadia are reported being those of Simmons and Yeargan (1988b) and Javahery (1990).

6.4 DESCRIPTIONS OF IMMATURE STAGES

The eggs (Deitz et al. 1976; Esselbaugh 1946; Miner 1966a; Sorenson and Anthon 1936; Underhill 1934; Whitmarsh 1914, 1917) and nymphal instars (DeCoursey and Esselbaugh 1962, Deitz et al. 1976, Miner 1966a, Morrill 1910, Underhill 1934, Whitmarsh 1917) have been described thoroughly and/or illustrated. Recently, Javahery (1994) discussed egg development in several pentatomoids, including *Acrosternum hilare*; he presented a diagram of an egg cluster of this bug in which he showed the orientation of the developing embryos.

6.5 ECONOMIC IMPORTANCE

6.5.1 Soybean

6.5.1.1 General Information

Acrosternum hilare, Nezara viridula, and *Euschistus servus* have been listed as the three most important members of the stink bug complex that attack soybean in North America (e.g., Funderburk et al. 1999, Todd 1976); Russin et al. (1987, 1988a, b) and Zeiss and Klubertanz (1994), among many others, have listed *A. hilare* as a major pest of this crop.

Jones and Sullivan (1982) studied the life cycle of *Acrosternum hilare* in South Carolina in relation to soybean. They considered the bug to be bivoltine. Early in the year, the bugs fed and reproduced on black cherry and elderberry, which allowed populations to increase. The resulting adults (summer generation) then moved from these hosts into soybean. In fact, black cherry and elderberry frequently grew along fence rows bordering soybean fields. These summer adults began entering soybean fields in mid-July; their numbers decreased during the second week of August, and large nymphs appeared 1 month later. Bug populations peaked in soybean in September and early October.

Miner (1966a, b) reported similar results for Arkansas. The first generation occurred on wild hosts, especially dogwood, and the second generation on soybean. Specifically, he found that overwintered adults laid their eggs in early June on dogwood, and the nymphs and adults fed on the berries until the berries matured and dropped. They also fed on other plants with berries or pods including elderberry, peppervine, wild grape, sumac, black locust (Miner 1966a, b), and pecan (Miner 1966a). The resulting adults appeared shortly after soybeans began setting. These adults laid their eggs on soybean leaves. The resulting nymphs became more numerous and peaked from mid-September to early October (Figure 6.1). Most damage to soybeans occurred from early September until near harvest time. Heaviest soybean infestations were consistently found on border rows next to woods, similar to the findings of Jones and Sullivan (1982). Nearness of the field to the preferred wild host plants seemed to be a major factor in determining the severity of the infestations.

McPherson et al. (1993) and Bundy and McPherson (2000b) found that, in Georgia, stink bug population densities in soybean, including those of *Acrosternum hilare*, began to increase steadily in mid-August as pods began to fill with seeds and

FIGURE 6.1 *Acrosternum hilare* nymphs on soybean pod. (Courtesy of R.M. McPherson, University of Georgia.)

peaked from late September to early October. Buschman et al. (1984) reported similar results for Mississippi. Raney and Yeargan (1977), based on a 4-year study conducted in Kentucky, found *A. hilare* was most abundant late in the season. This was not surprising because it feeds on soybean pods and developing seeds rather than foliage. Bickenstaff and Huggans (1962) reported that *A. hilare* was most common in soybean fields in Missouri later in the year.

6.5.1.2 Damage

McPherson et al. (1979a) reported that *Acrosternum hilare* can cause damage to soybean yield and quality, although the severity of this attack is less than that of *Nezara viridula*. They found that fifth instars and adults caused significantly less damage than the same stages of *N. viridula*, but that fourth and fifth instars and adults caused as much or more damage than fourth instars of *N. viridula*. Plants exposed during the R5 and R6 stages of development were affected most seriously in quality and yield losses. Significantly, however, they found that the bug's populations did not increase abruptly during these plant developmental stages and, thus, its numbers seldom reached an economic threshold. Jones and Sullivan (1983) exposed soybeans during the R4 and R5 developmental stages to densities of 1 or 3 caged adults per 0.46 m (18 in.) of row of *A. hilare*, *N. viridula*, and *Euschistus servus* and found no significant difference in the type and amount of feeding damage on soybean seed by the three species.

As one moves northward, the abundance of *Acrosternum hilare* increases relative to that of *Nezara viridula* (McPherson et al. 1994, Turnipseed and Kogan 1976). For example, Jones and Sullivan (1978) reported that during their field testing in South Carolina in 1974, *N. viridula*, *A. hilare*, and *Euschistus* spp. comprised 60, 25, and 14% of the infestation, respectively; and in 1975, 68, 17, and 14%, respectively; similarly, in 1979, they reported from their 1973 field testing percentages of 60, 19, and 20%, respectively, for these same taxa. Miner (1966a, b) reported that *A. hilare* was responsible for most soybean damage in Arkansas. *N. viridula* is not listed for Kentucky (Froeschner 1988) and its presence in Illinois is based on a single specimen (McPherson and Cuda 1974), but *A. hilare* is common in both states. Raney and Yeargan (1977) found that *A. hilare* attacked soybean pods and developing seeds in Kentucky. Based on all insects collected during their 4-year study, they concluded that Kentucky soybean insect populations are more similar to those of the Midwest than to those of the Deep South.

Yeargan (1977) found that *Acrosternum hilare* caused significant reductions in yield and percent germination when the plants were exposed to densities of 2 or 4 adults/0.3 m (1 ft) of row from early pod development to harvest maturity. Yield loss resulted from both reduced seed size and decreased numbers of seeds produced. Bug densities of 1, 2, or 4 bugs/0.3 m of row caused significant increases in the percentage of beans damaged. He also found that the fifth instars damaged a higher percentage of beans than any other stage during a 1-week period. Subsequently, Simmons and Yeargan (1988a) found that approximately one half of the total number of nymphal feeding episodes occurred during the fifth stadium, and that fifth instars had the highest daily frequency of feeding (3.4 times/day). Interestingly, adult body size was not related to the total number of times a bug fed during nymphal development.

Miner (1966a, b) reported that an infestation of 1 bug/3 ft of row caused enough damage to justify insecticidal treatment. He also found that soybeans that had been damaged but were essentially full-sized had suffered approximately 9% weight loss. Early infestations by high numbers of bugs resulted in marked reduction of soybean size. He stated there was much evidence that the longer the pods were exposed to high numbers of the bugs (i.e., in the fall), the greater the damage. Therefore, late-maturing varieties suffered the most damage. It is important to note, however, that because the bug's populations are usually high late in the growing season when beans are full-sized or nearly so, the damage they cause is usually light.

Bickenstaff and Huggans (1962) conducted field and laboratory tests on the feeding damage of *Acrosternum hilare* to soybean. In field studies, they exposed caged plants to densities of 5 and 10 adults and noted reproduction of the bugs only at the higher density. Exposing plants to 10 adults initially (and, apparently, their offspring) resulted in decreased yield, decreased number of seeds harvested, increased percentage of small seeds, increased percentage of discolored and moldy seeds after storage, increased number of pods with no seeds or only one seed, and decreased stem length. They determined, in the laboratory, that when the bugs were confined on soybean seedlings, they fed readily on leaf stems and main stems and caused shriveled leaves and stem dieback.

Simmons and Yeargan (1990) studied the separate and combined effects of manual defoliation of the main stem of soybean beginning at growth stage R1 (beginning bloom) and feeding of *Acrosternum hilare* on yield and seed quality. Branches responded to defoliation by producing an additional 10 to 15% foliage by the end of the defoliation period (R2 developmental stage, full bloom). By the R5 growth stage (beginning of seed production), approximately 60% of the plant's foliage was on the branches. Injury from feeding by *A. hilare* (infestation levels up to 3.0 bugs/0.3 m of row) resulted in reduction in yield and seed quality. Yield reduction, which increased with infestation level, resulted from lower numbers of seeds. Also, as infestation level increased, percentage of damaged seeds, number of seeds fed upon, and number of times seeds were fed upon increased; and vigor and viability decreased.

Miner and Wilson (1966) and Miner and Dumas (1980a, b) reported that fat acidity was significantly higher in soybeans stored at high moisture levels than at low levels but (Miner and Dumas 1980a, b) only when before- and after-storage data were combined. Although Miner and Wilson (1966) stated that the greatest increase in fat acidity was in soybeans that were both heavily damaged and stored in high moisture, Miner and Dumas (1980b) found no relationship between acidity and damage.

Miner and Wilson (1966) reported that before storage, heavily damaged soybeans averaged slightly less oil and slightly more protein than slightly damaged beans. Miner (1966a, b) reported that damaged soybeans averaged approximately 1% less oil and 1% more protein than undamaged beans. Daugherty et al. (1964) also reported less oil and more protein in damaged beans; the difference was more dramatic in beans planted on the earliest date of the experiment (April 19) than on the later dates, collectively (May 24, May 31, June 19). They also found a significant decrease in the size of damaged beans compared to undamaged beans. Miner and Dumas (1980b) reported that damaged soybeans had more split seeds and broken seed coats than undamaged ones.

Daugherty et al. (1964) tested feeding by *Acrosternum hilare* on five varieties of soybean planted on four different dates. They found the longer the period from bloom to maturation, the more severe the damage; they suggested that this was because faster maturing plants were available for feeding for shorter periods of time.

Finally, several investigators have shown that *Acrosternum hilare* is capable of transmitting *Nematospora coryli*, which causes yeast-spot disease in beans (Clarke and Wilde 1970a, b, 1971; Daugherty 1967; Foster and Daugherty 1969; Leach and Clulo 1943).

6.5.2 CORN

6.5.2.1 Damage

Acrosternum hilare attacks corn (Mundinger and Chapman 1932, Swan and Papp 1972, Tonhasca and Stinner 1991) and, potentially, is capable of causing economic damage. However, Townsend and Sedlacek (1986), from their greenhouse studies, found that damage to seedling corn caused by the feeding of this stink bug was much less severe than that caused by *Euschistus servus* and *E. variolarius*. They suggested that this might not be due to a difference in the ability of *A. hilare* and *Euschistus* spp. to cause damage to this growth stage but to a difference in propensity to feed on seedling corn. Mean dry root weights of plants exposed at 0.5 stage corn where bugs were confined to the base of the plant were significantly less than the mean control weight. Finally, tillering was exhibited by 30% of the exposed plants.

6.5.3 TOBACCO

6.5.3.1 Damage

Acrosternum hilare now is listed as a pest of tobacco (Reich 1991). Adults and nymphs extract juices from all parts of the plant but prefer more tender growth. This may result in the leaves wilting and flopping over.

6.5.4 FRUITS

6.5.4.1 Damage

Acrosternum hilare long has been known to attack the developing fruits of various cultivated trees including, particularly, peach (e.g., Butler and Werner 1960; Chandler 1950; Chandler and Flint 1939; Phillips 1951; Porter et al. 1928; Rings 1955, 1957; Snapp 1941, 1954; Sorenson and Anthon 1936; Swan and Papp 1972; Underhill 1934; Whitmarsh 1914, 1917; Woodside 1946b, 1950) and pear (Mundinger and Chapman 1932, Phillips 1951, Sorenson and Anthon 1936). However, most studies have dealt with peach. Here, and probably with most other fruits, the type of damage varies with the age of the developing fruit. The earlier in the development of the fruit, the more severe the damage. The more severe damage is called catfacing and is characterized by sunken areas surrounded by tissues that are more or less distorted (Porter et al. 1928); the surfaces are rough, corky, and lack pubescence (Woodside 1946b, 1950). The sunken areas occur because the tissues at the site of the feeding puncture cease to grow, whereas those surrounding the site do not (Whitmarsh 1914, 1917).

Dimpling (or scarring) is a slight depression (Porter et al. 1928) caused by shrinkage of the tissues injured by the feeding of the bugs (Woodside 1946b), but, unlike catfacing, the pubescence persists (Woodside 1946b, 1950); it is considered an atypical type of catfacing (Rings 1957). Rings (1957) found that catfacing occurred when the developing fruits were attacked from bloom up to 49 days after bloom, and scarring (or dimpling) occurred from 42 to 56 days after bloom. Chandler (1950) and Chandler and Flint (1939) reported that *Acrosternum hilare* specializes in dimpling of peaches, and Porter et al. (1928) noted the nymphs sometimes cause considerable damage in the form of small dimples. Woodside (1950) stated that this species caused most dimpling observed in Virginia orchards and that both nymphs and adults fed on the developing fruits. Feeding damage to peaches is discussed further by Sorenson and Anthon (1936) from their work in Utah orchards.

Reports of damage to pear have been less frequent but, apparently, can be as severe. Mundinger and Chapman (1932) reported that an orchard was so heavily infested with *Acrosternum hilare* that one jarring of a tree brought down scores of bugs, mostly adults. They also observed this bug feeding on apple. They found that the most severe damage to fruit occurred in the vicinity of woodlands or where the orchard bordered uncultivated land. The bugs seemed most attracted to green (developing) pear fruits. Sorenson and Anthon (1936) reported that the damage to pears caused by feeding resulted in pits or depressions over the surface surrounding the punctures.

The most comprehensive study of the relationship of *Acrosternum hilare* to developing fruits is that of Sorenson and Anthon (1936) in Utah in which they followed the movement of the bugs among different species of fruit trees as the fruits developed, again showing the typical behavior of this stink bug to use several different hosts during the season. Between June 15 and July 1, most of the bugs were found on developing cherries, which developed small depressions or pits surrounding the feeding wounds. After the cherries were harvested, the bugs moved to ripening apricots, which were harvested by early August. Then the bugs moved to early varieties of peaches and, by August 20, it was possible to find two or three bugs on almost every peach. Toward the end of the peach season, the bugs moved to ripening pears. Obviously, this species had the potential to do much damage to these fruits; apparently damage to peach was most severe on early varieties (Sorenson and Anthon 1936).

6.5.5 COTTON

6.5.5.1 General Information

Acrosternum hilare attacks cotton and is capable of severely damaging the bolls. Barbour et al. (1988) reported that it was the predominant phytophagous stink bug in cotton. Turnipseed et al. (1995) and Bundy and McPherson (2000b) noted during their studies that it and other stink bugs were major secondary pests of cotton.

Barbour et al. (1988) found that in North Carolina the adults occurred in cotton initially about mid-June, increased in numbers by mid-July, and peaked in early August. Nymphs were found soon after August 1 and became the main stage by mid-August. Movement into the cotton fields occurred primarily when the plants were setting bolls, with most bolls being 2-weeks-old or less.

6.5.5.2 Damage

Barbour et al. (1988), from field-cage studies, found the number of punctures per boll increased significantly as exposure time to the bugs increased, and percent harvestable locks decreased significantly. Yield loss averaged 25% from a 6-day exposure at a density of 3 adult bugs/plant and as high as 66% yield loss from a 24-day exposure at the same density. Also, from greenhouse studies, they found that third to fifth instars and adults significantly reduced the number of harvestable locks, with the reduction caused by fifth instars equal to that of the adults. Barbour et al. (1990) found a lower seed germination rate and an increase in percentage of immature fiber of lint harvested from cotton bolls previously exposed to the bugs. Roach (1988), using 0, 1, or 2 adults in field cages containing individual cotton plants, reported that the bugs appeared to have reduced the number of harvestable locks remaining on the plants, but the results had been obscured by weather factors such as drought. Turnipseed et al. (1995) reported that *Acrosternum hilare*, *Nezara viridula*, and *Euschistus servus* were the major secondary pests observed in untreated

Bt cotton during their study of Bt cotton in South Carolina. They found 7% damaged bolls in non-Bt cotton treated twice with Karate® insecticide compared with 27% damaged bolls in untreated Bt cotton.

Granted the above information on the damage caused by *Acrosternum hilare* is not surprising, but what is fascinating is why this species and other species of stink bugs (e.g., *Euschistus servus*) seem to be increasing in importance as cotton pests. Apparently, it is related directly, at least in part, to the current control efforts against the boll weevil (Bacheler and Mott 1995, Lambert and Herzog 1993). In addition, it is possible that other species of stink bugs that are considered today to be minor pests, but which were major pests before the invasion of the boll weevil into the United States, may well become major pests again.

Acrosternum hilare and several other heteropteran species have been reported as pests of cotton since the mid-1800s (see Morrill 1910). However, it was the invasion of the boll weevil in the 1920s that devastated cotton production in the United States and precipitated the collapse of the cotton industry. For example, Leiby (1928) reported that although 1927 was a generally favorable growing season for cotton, the survival of the weevils during the previous winter had been above average, and 1927 also was favorable for their development. As a result, the weevils caused significant damage throughout the cotton states; of the states surveyed (Alabama, Oklahoma, and Tennessee were not included), the weevils accounted for approximately a 21% reduction of the 9,854,000 bales forecasted the previous December by the U.S. Department of Agriculture.

Barbour et al. (1988) discussed the entomological changes that have occurred in the North Carolina cotton field agroecosystem since the 1920s, but these changes are not limited to North Carolina. They noted that with the introduction of organic insecticides in the late 1940s and early 1950s, it once again became possible to control cotton pests and, apparently, to have stable production. Although cotton yields increased over the next 3 decades from the use of organic insecticides, the industry struggled with controlling the boll weevil adequately, with increased losses from the bollworm complex, and with persistent crop maturity delays associated with frequent use of organophosphate insecticides. A side benefit from use of insecticides was the control of stink bugs.

Implementation of the Boll Weevil Eradication Trial Program in the late 1970s (see Ganyard et al. 1979) and introduction of synthetic pyrethroids created the possibility of farmers producing cotton without the boll weevil posing a major threat. Also, pyrethroid insecticides had been shown to give excellent control of the boll-worm and good control of stink bugs.

As the use of insecticides has dropped during the 1980s and 1990s, insects such as stink bugs increasingly have begun to utilize this once again vulnerable crop. Among the more important of these are *Acrosternum hilare* and *Euschistus servus*. It will be interesting to see what the future holds concerning the economic impact of these species.

6.5.6   OTHER CROPS

6.5.6.1   Damage

Examples of other cultivated crops attacked by *Acrosternum hilare* include orange, okra, pea, bean, tomato, turnip, cabbage, eggplant (Swan and Papp 1972), lima bean (Schoene and Underhill 1933), and pecan (Smith 1996a, b). Schoene and Underhill (1933) noted that when lima beans were attacked, yeast-spot disease, caused by *Nematospora phaseoli* Wingard, greatly increased and generally was associated with the feeding of this stink bug.

7 *Oebalus* spp.

CONTENTS

7.1 *OEBALUS PUGNAX PUGNAX* (FAB.)

7.1.1 Introduction

Oebalus p. pugnax, the rice stink bug, ranges from New York and Connecticut south to Florida and west to Nebraska, Colorado, and Arizona (Froeschner 1988). First reported from rice, *Oryza sativa* (L.), in the 1880s by Riley (1882), it now is considered a major pest of this crop (e.g., Brook 1953, Douglas 1939, Douglas and Ingram 1942, McPherson 1982, Sailer 1944, Odglen and Warren 1962, Swanson and Newsom 1962, Way 1990, Webb 1920). Although it feeds on a wide range of other cultivated crops and wild hosts, it shows a distinct preference for grasses (e.g., Johnson et al. 1987, Odglen and Warren 1962). Hosts, in addition to rice, include small grain (Anonymous 1973), sorghum (Dahms 1942; Franqui et al. 1988; Hall and Teetes 1980, 1981, 1982a, c; Hall et al. 1983; Harper et al. 1993, 1994; Odglen and Warren 1962; Sailer 1944), wheat (Forbes 1905; Lugger 1900; Odglen and Warren 1962; Sailer 1944, Viator and Smith 1980; Viator et al. 1982, 1983), barley, rye, oats, corn (Odglen and Warren 1962), *Sorghum halepense* (L.) (Davis 1925, Franqui et al. 1988, Hall and Teetes 1981, McPherson and Mohlenbrock 1976, Odglen and Warren 1962), *Setaria* (Froeschner 1941, Garman 1891), *Panicum dichotomiflorum* Michaux, *Digitaria sanguinalis* (L.), *Echinochloa colona* (L.), (Douglas 1939, Douglas and Ingram 1942), *Paspalum urvillei* Steudel (Douglas 1939, Douglas and Ingram 1942, Hall and Teetes 1981), *Echinochloa crusgalli* (L.) (Douglas and Ingram 1942, Johnson et al. 1987, Lee et al. 1993, Odglen and Warren 1962), *Panicum fasciculatum* Swartz, *Paspalum longipilum* Nash (Douglas 1939), *Cynodon dactylon* (L.) (Douglas 1939, Hall and Teetes 1981), *Lolium perenne* L., *Lolium multiflorum* Lamarck (Lee et al. 1993), *Phalaris minor* Retzius (Douglas and Ingram 1942), *Paspalum dilatatum* Poiret (Hall and Teetes 1981, Odglen and Warren 1962), *Paspalum distichum* L., *Panicum hians* Elliott, *Glyceria septentrionalis* Hitchcock, *Carex* spp., *Rhynchospora inexpansa* (Michaux) (Odglen and Warren 1962), *Avena sativa* L., *Brachiaria fasciculata* (Swartz), *Brachiaria texana* Buckley, *Hordeum vulgare* L., *Paspalum pubiflorum* Ruprecht, *Panicum coloratum* L. (Hall and Teetes 1981), and others. *O. p. pugnax* also has been reported as a predator of the "cotton worm" [*Alabama argillacea* (Hübner)] (Riley 1885).

7.1.2 Life History

Oebalus p. pugnax overwinters as adults in grassy areas (Douglas 1939, Douglas and Ingram 1942, Drees 1983, Grigarick 1984, Hamer and Jarratt 1983, Ingram 1927, Nilakhe 1976a, Pathak 1968, Texas Agric. Ext. Serv. 1997), trash (Douglas 1939, Hamer and Jarratt 1983, Ingram 1927, Johnson et al. 1987, Odglen and Warren 1962, Pathak 1968, Rolston and Kendrick 1961), and ground litter (Way 1990). Adults begin entering overwintering sites in early October in Louisiana (Nilakhe 1976a) but become active when disturbed (Douglas 1939). They emerge in spring (Bowling 1967, Douglas 1939, Drees 1983, Grigarick 1984, Ingram 1927, Nilakhe 1976a, Pathak 1968, Texas Agric. Ext. Serv. 1997), males preceding females by approximately 10 days (Nilakhe 1976a); overwintering females are unmated and contain hypertrophied fat bodies (Nilakhe 1976a). They begin feeding on wild grasses (Bowling 1967, Douglas 1939, Douglas and Ingram 1942, Grigarick 1984, Ingram 1927, Pathak 1968), almost exclusively on the developing seeds (Bowling 1967, Douglas and Ingram 1942, Grigarick 1984, Ingram 1927, Pathak 1968). Two of the most important of these grasses are *Echinochloa crusgalli* (barnyard grass) (Odglen and Warren 1962) and *Paspalum urvillei* (vasey grass) (Douglas 1939, Douglas and Ingram 1942). Reproduction begins shortly thereafter. *O. p. pugnax* is apparently bi- or trivoltine (McPherson and Mohlenbrock 1976, Odglen and Warren 1962). Reports of four or more generations per year (Douglas 1939, Douglas and Ingram 1942, Ingram 1927, Smith et al. 1986) need confirmation.

7.1.2.1 Mating Behavior

Precopulatory and copulatory behavior of *Oebalus p. pugnax* has been described in detail by Nilakhe (1976a). Typically, a male approaches a female with his antennae vibrating and genitalia extended. He strokes her antennae several times and walks around and over her while continually stroking her with his antennae. He then raises the end of her abdomen with his head 10 to 15 times, forcing it upward, and quickly rotates his body so that his genitalia can come into contact with hers. If she has lowered her abdomen, he will use his hind legs to elevate it. If still unsuccessful, he again will raise the end of her abdomen with his head until she keeps it elevated so copulation can occur. If the female is receptive, coupling can occur within 5 minutes. If unreceptive, the female will lower her abdomen, kick at the male with her hind legs, or walk away. Mating occurs during the day (McPherson and Mohlenbrock 1976), evening (Nilakhe 1976a), and night (Davis 1925).

The ability of *Oebalus p. pugnax* to feed and reproduce on a wide range of wild grasses plays a significant role in its status as an economic pest. When adults become active in the spring, only wild grasses are available as suitable hosts. Therefore, early reproduction occurs on these hosts, resulting in increased numbers of bugs. The subsequent impact of this stink bug on cultivated plants increases in severity the later in the year those plants become suitable as hosts.

7.1.2.2 Effects of Abiotic Factors on Populations

Mortality factors differ in importance for adults and nymphs. Adult numbers are reduced by severe winter conditions and nymphal numbers by high summer temperatures (Douglas and Ingram 1942, Ingram 1927, Pathak 1968).

7.1.3 LABORATORY INVESTIGATIONS

This stink bug has been reared under confined conditions (Esselbaugh 1948, Ingram 1927, Nilakhe 1976a, Odglen and Warren 1962); attempts to rear it on a meridic Debolt diet have been unsuccessful (Brewer and Jones 1985). Nilakhe (1976a) reported that females reared on rice oviposited an average of 915 eggs, approximately twice that of bugs reared on vasey grass and barnyard grass. Naresh and Smith (1983) reared this bug on rice, sorghum, and vasey grass at 21, 24, 27, and 30°C and generally found no difference in relative weights of nymphs; however, relative weights of adults reared on the three hosts differed, depending on temperature.

Survival of nymphs generally was greater on rice and grain sorghum than on vasey grass. Mean developmental time (egg to adult) was shortest on sorghum at 21 and 27°C and not significantly different between the three plants at 24 and 30°C. Naresh and Smith (1984) also reported on feeding preference of *Oebalus p. pugnax* between vasey grass, 11 other species of grasses, and sedge (*Cyperus*). The bugs preferred panicles of vasey grass over those of the other plants, with the number of feeding bugs doubling from 1 hr to 6 hr postinfestation. When given a choice between panicles of cultivated rice and vasey grass, the bugs fed more on rice at 6 hr postinfestation. When given a choice between different panicle lengths (i.e., 7, 14, 21, and 28 cm) of these same plants, they fed more on the 21- and 28-cm long rice panicles than on the 7-cm long panicles but showed no preference between panicle lengths of vasey grass.

7.1.4 DESCRIPTIONS OF IMMATURE STAGES

The eggs (Douglas and Ingram 1942, Esselbaugh 1946, Garman 1891, Ingram 1927, Odglen and Warren 1962, Smith et al. 1986) (Figures 7.1 and 7.2) and various instars (DeCoursey and Esselbaugh 1962, Douglas and Ingram 1942, Ingram 1927, Odglen and Warren 1962, Smith et al. 1986) have been described and/or illustrated.

7.1.5 ECONOMIC IMPORTANCE

7.1.5.1 Rice

7.1.5.1.1 General Information
Oebalus p. pugnax attacks several cultivated crops, but its damage to rice has received the most attention in recent years. It is considered the most important pest of rice during grain formation in the United States (Grigarick 1984). The seriousness of its infestation is related to the ratio of the bugs to grain, stability of the bug populations, effect of parasites on the eggs, presence or absence of various fungi, and the shapes of the grain (Bowling 1967). Its economic impact can be substantial.

Oebalus p. pugnax overwinters as adults, which emerge in early spring to feed and reproduce on wild grasses, particularly those that are producing seed (e.g., Douglas and Ingram 1942). Their offspring reproduce on wild grasses or on rice or both. Harper et al. (1993, 1994) noted that adults move from alternate hosts (including weeds and grain sorghum) to rice when the heads are developing. Nymphal populations do not develop in rice fields until the grain begins to develop unless the fields and surrounding areas contain alternate weed hosts that mature earlier than rice.

There is some variation in the life cycle, depending on the geographical location of the insect. For example, in Arkansas, the bugs appear early on oats, barley, wheat, and wild grasses, being attracted to those plants that are producing seed (Odglen and Warren 1962). They move from one patch of heading grass to another as grass matures and dries out. Barnyard grass is particularly important at this time (Johnson et al. 1987, Odglen and Warren 1962). Population numbers increase on these grasses until rice begins to bloom during the summer, at which time the bugs gradually move from these grasses to rice (Odglen and Warren 1962).

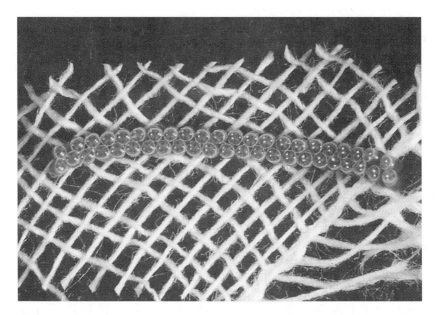

FIGURE 7.1 *Oebalus p. pugnax* mature egg mass (6.4×). (Courtesy of C.S. Bundy, University of Georgia.)

FIGURE 7.2 *Oebalus p. pugnax* mature egg mass (25×) with eye spots visible (see arrows) and rings of micropylar processes. (Courtesy of C.S. Bundy, University of Georgia.)

In Louisiana and Texas, the bug also feeds on a wide variety of grasses in the spring (Douglas 1939, Ingram 1927), preferring those grasses producing seed (Bowling 1967, Douglas and Ingram 1942, Drees 1983, Ingram 1927, Texas Agric. Ext. Serv. 1997). Of these, the most preferred wild host is vasey grass (Douglas 1939, Douglas and Ingram 1942). This grass begins to mature in early May but continues to produce panicles until late fall, thus providing the bugs with suitable food for at least 5 months (Douglas and Ingram 1942). Douglas (1939) stated that the bugs had not been observed feeding on any cultivated crop other than rice. It does well on rice field levees and fence rows, and along highways (Douglas 1939). When the rice panicles emerge (the most common variety heads-up approximately August 15 to 20), the bugs migrate from the grasses to rice, adults by flying and nymphs by crawling or swimming (Douglas 1939). Movement to rice is initiated when other hosts have diminished in abundance or are no longer suitable as food (Bowling 1967, Harper et al. 1993). Mowing grass along highway ditches will cause a sudden increase in bugs in adjoining fields. Cutting rice will stimulate the bugs to move to later-maturing fields (Douglas 1939).

Swimming by stink bug nymphs is unusual but not surprising for *Oebalus p. pugnax*, considering its preference for rice. According to Douglas (1939), the nymphs are active swimmers. When dislodged from a rice plant, they will swim to the nearest plant, using jerky movements, and climb to the panicle. If disturbed, they will drop into the water.

Jones and Cherry (1984) noted that in south Florida *Oebalus p. pugnax* begins to appear in fields as rice begins to head, no matter the calendar date. Similarly, Hamer and Jarratt (1983) noted that in Mississippi the bugs begin to migrate to rice soon after the plants begin to head.

Douglas (1939) reported the average population in rice fields was approximately 750 bugs/acre from 1 to 3 days after emergence of the panicles. Females oviposited on the leaves, stems, and panicles. The resulting population increased at a rate of approximately 1,000/week as the rice was maturing and, in fact, all stages of the bugs could be found during this time. Maximum populations usually did not develop until a few days after the rice was in an optimum condition for feeding because this time had past before some of the nymphs developing from eggs laid within the fields had begun feeding (Douglas 1939). When the rice kernels became too hard for feeding, the bugs migrated back to grasses and levees before entering overwintering sites. Webb (1920) found that injury to rice by *Oebalus p. pugnax* occurred primarily in late summer and early fall when the heads were developing, the damage resulting from attack of adults and late instars.

Odglen and Warren (1962) studied *Oebalus p. pugnax* in Arkansas, limiting their study to fields (i.e., they did not include the levees surrounding these fields). They found a definite association between the amount of barnyard grass in rice fields and the number of bugs. Practically no bugs were found in areas of the fields with lesser amounts of grass between early June and mid-August. Numbers increased dramatically in grassy sites during this period, peaking in early August; they remained high during August, when rice was most susceptible to damage. However, after early August, numbers of bugs in the grassier areas began to decline but not because the bug population was decreasing. Actually, there was a buildup of numbers overall.

The decline in grassy areas was because the bugs began moving from grass to rice as the rice began to emerge from the boot. Therefore, the bugs were more evenly dispersed throughout the field, resulting in fewer bugs per sample. By the dough and "hard grain" stages of rice development, the bugs were distributed almost uniformly over the fields. Numbers of bugs remained low throughout the season in sites free of grass, even after heading. These results indicated that grassy sites were the source of infestation for all parts of the field and, further, implied that eliminating wild grasses from the field would reduce damage to rice from this stink bug.

Douglas (1939) also studied this bug in rice fields but included levees, particularly those associated with rice fields in Texas. Surprisingly, he stated that there was no relationship between abundance of the bugs in rice fields and the abundance of wild hosts on levees. This would seem to contradict, at least on the surface, the results of Odglen and Warren (1962) (i.e., the more grass present, the higher the population of bugs on rice). Douglas (1939), based on 400 sweeps/field, reported that for fields with wild hosts on the levees associated with them, the average number of bugs per acre was 3,465 for the edge of the field and 3,133 for the center; in fields without wild hosts, the average number of bugs per acre was 1,585 for the edge of the field and 2,448 for the center. These data would appear to support Odglen and Warren (1962), but Douglas (1939) stated that the higher averages for fields with wild hosts were because of an extremely high population of bugs in one of those fields. To complicate matters, Ingram (1927) noted that rice fields (geographic location not given but probably Louisiana and/or Texas) associated with much grass invariably suffered heavier losses, apparently contradicting Douglas (1939). Finally, Douglas and Tullis (1950), in their Louisiana study, sampled the bug populations on levees adjacent to rice fields, in the fields near the levees, and in the centers of the fields from the time the panicles emerged until they matured (dates not given). They found little change in numbers on the levees. However, there was a dramatic increase within the fields, particularly near the levees, until approximately the twelfth day when the kernels reach the dough stage, and then there was a decrease.

Adults begin entering overwintering sites in early October in Louisiana (Nilakhe 1976a). However, they can be on rice for a much longer time if a ratoon (second) crop is grown. In southern Florida, where two crops can be grown during the year, rice is planted as early as early March (Jones and Cherry 1984). The bugs have been found there continuously from June 10 to November 10 (Jones and Cherry 1986); in fact, Green et al. (1954) stated that they still were common in plantings until mid-October. Jones and Cherry (1986) reported that heading of rice in southern Florida first occurred in mid- to late June, and that the bug populations peaked in late July to early August, corresponding to when the rice fields were fully headed. Population levels also were high in October and November in the ratoon crop, both because this second crop was heading at this time and because insecticides were not used in this crop. Rapid population increases can result from adults migrating into the fields, hatching of eggs after insecticide application, or both. Foster et al. (1986) and Jones and Cherry (1984) also reported rapid increases in population density at heading, peaking during grain-filling in both crops. Adults were the most common stage during the first crop and nymphs the most common stage during the ratoon crop (Foster et al. 1989).

7.1.5.1.2   Number of Generations Annually

The reported number of generations/year has differed among authors, perhaps because of different geographic locations of the studies. Odglen and Warren (1962) reported that, in Arkansas, overwintered adults produced at least a partial brood on early grasses and then migrated directly to rice in early June. First-generation adults appeared in rice in late June, overlapping broods occurred in July and August, "with two generations apparently occurring in this area." Therefore, they reported, there are evidently two or three generations/year. Ingram (1927) and Douglas and Ingram (1942) stated that, presumably in Louisiana and Texas, *Oebalus p. pugnax* emerged from overwintering sites in late April or early May, began feeding and reproducing on a large variety of grasses, and passed through two or three generations. Subsequently, it moved to rice as this plant began heading up and passed through two or three generations. Finally, they noted that some bugs continued to feed on grasses throughout the season, producing four or five generations/year. Therefore, they felt, apparently, that this stink bug had four to six generations/year. It is more difficult to understand what Douglas (1939) was trying to say. He reported that the bug passed through two or three overlapping generations on wild hosts and one complete and three partial generations in rice fields by the time the rice was harvested (we are assuming that Douglas meant the rice was harvested when nymphs were present because, obviously, nymphs do not reproduce). Grigarick (1984), in his general discussion of this bug, stated that it fed on developing seeds of a wide range of grasses and passed through several generations before rice began heading. However, unlike the above authors, he stated that it had only a single generation on rice within an individual field.

7.1.5.1.3   Damage

Oebalus p. pugnax is attracted to rice, primarily during the reproductive phase of growth, particularly during the grain-filling period. Grain filling is subdivided into the milk, soft- and hard-dough stages, the terms based on the liquid content within the developing grains. Both adults and nymphs feed on the developing grains (Bowling 1967, Douglas 1939, Douglas and Ingram 1942, Douglas and Tullis 1950, Gifford et al. 1968b, Grigarick 1984, Ingram 1927, Pathak 1968, Swan and Papp 1972, Webb 1920). Rolston et al. (1966) noted the bugs showed a definite preference for grain in the earlier stages of plant development, suggesting that the bugs could cause measurable damage by destroying younger kernels or by arresting the development of somewhat older kernels.

Feeding results in yield loss and/or reduced quality (Drees 1983, Elliott et al. 1994, Smith et al. 1986, Swanson 1960, Swanson and Newsom 1962, Texas Agric. Ext. Serv. 1997), depending on when the grain is attacked during development (Chambliss 1920, Gifford et al. 1968b, Harper et al. 1993). When attacked during the milk stage, the contents are sucked out, resulting in false grains (i.e., empty glumes, empty seed coats) (Bowling 1956, 1967; Douglas and Ingram 1942; Drees 1983; Genung et al. 1979; Gifford et al. 1968b; Hamer and Jarratt 1983; Ingram 1927; Johnson et al. 1987; Odglen and Warren 1962; Swan and Papp 1972; Texas Agric. Ext. Serv. 1997) and atrophied grains (Bowling 1967, Chambliss 1920), which result in yield loss (Bowling 1956, Gifford et al. 1968b, Smith et al. 1986, Swanson and Newsom 1962). When attacked during the soft- and hard-dough stages, only a portion of the contents is removed, leaving a chalky discolored area around the feeding site (Douglas and Ingram 1942, Gifford et al. 1968b, Ingram 1927, Johnson et al. 1987, Odglen and Warren 1962, Swanson and Newsom 1962). Rice so damaged is called "pecky" rice (Douglas and Ingram 1942, Genung et al. 1979, Gifford et al. 1968b, Ingram 1927, Johnson et al. 1987, Odglen and Warren 1962, Swan and Papp 1972, Swanson and Newsom 1962) (Figure 7.3); although not as common, peck also is found in rice attacked during the milk stage (Harper et al. 1993, Ito 1978). Pecky rice is of reduced quality (Bowling 1956, 1963; Daugherty and Foster 1966; Douglas 1939; Elliott et al. 1994; Genung et al. 1979; Gifford et al. 1968b; Harper et al. 1993, 1994; Smith et al. 1986; Swanson and Newsom 1962; Way 1990) because it is weakened in the region of the feeding site and often breaks during milling (Bowling 1956, 1967; Chambliss 1920; Daugherty and Foster 1966; Douglas and Ingram 1942; Douglas and Tullis 1950; Elliott et al. 1994; Genung et al. 1979; Gifford et al. 1968b; Ito 1978; Johnson et al. 1987; Harper et al. 1993, 1994; Odglen and Warren 1962; Swanson and Newsom 1962; Way 1990) and because of discoloration associated with feeding (Bowling 1967; Chambliss 1920; Douglas and Tullis 1950; Elliott et al. 1994; Gifford et al. 1968b; Harper et al. 1993, 1994; Ito 1978; Smith et al. 1986; Swanson and Newsom 1962). If pecky rice does not break during milling, it will appear in head rice and, thus, the rice will be of inferior quality because of the discoloration (Bowling 1967, Ito 1978, Swanson and Newsom 1962).

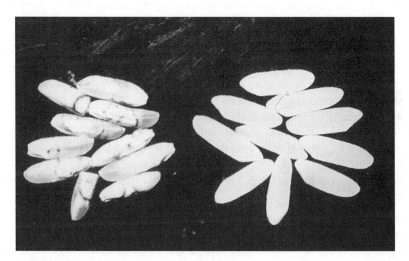

FIGURE 7.3 Pecky rice (left) resulting from stink bug feeding, and undamaged rice (right). (Courtesy of M.O. Way, Texas A&M University.)

7.1.5.1.4   Pecky Rice: What Is It?

Although the term "pecky rice" continually appears in the rice literature, there apparently is no standard definition of what pecky rice is. In fact, a search of the literature shows the description of the appearance and cause of pecky rice has changed over the last 60 years.

Ingram (1927) stated that feeding by this stink bug during the dough stage left a chalky, discolored area around the feeding site and that this damaged rice was called pecky. Pecky rice was described similarly by several subsequent authors including Douglas and Ingram (1942), Gifford et al. (1968b), Johnson et al. (1987), Odglen and Warren (1962), and Swanson and Newsom (1962). Ingram (1927) noted that a fungus could enter the grain through the feeding punctures and cause a black speck. Douglas and Ingram (1942) noted that the spots (= discolored areas of Ingram [1927]) could vary from light yellow to black; that the most common spot was circular and approximately 2 mm in diam. (ranging from the size of a pin point to general discoloration of the kernel); that, according to rice pathologists, similar discoloration sometimes was caused by fungi; and that the size of the spot could be affected by fungi entering the damaged kernel, causing further discoloration. Today, the definition of pecky rice has expanded from a description of kernels with spots (Lee and Tugwell 1980) to a convenient catchall expression for various types of rice kernel imperfections (Lee and Tugwell 1980, Tugwell 1986).

Some of the expanded definitions of pecky include the following: Douglas and Tullis (1950) stated that pecky rice was a term for kernels with spots, which were injuries caused by stink bugs or disease organisms. Bowling (1967) stated that grain also was fed upon in the later dough stage and, if infected with certain fungi, developed into grain with a black spot (black speck of Ingram [1927]?) and, commonly, was termed pecky rice. Drees (1983; also see Texas Agric. Ext. Serv. 1997) stated that pecky rice was shriveled kernels with spots varying from light yellow to black, which lowered the market value. Hamer and Jarratt (1983) and Johnson et al. (1987) indicated that peckiness could be caused by factors other than stink bug feeding. Harper et al. (1993) stated that feeding by this bug in the dough stage was thought responsible for reduced quality by introducing fungi into rice kernels, resulting in "pecky" rice. Gifford et al. (1968b) stated that the term "pecky" designated "kernels having spots caused by stink bug feeding punctures, and/or diseases which structurally weaken and cause breakage of the kernel, or discoloration." And Smith et al. (1986) stated that lesions on kernels resulting from feeding during the dough stage were invaded by several secondary fungi, which caused chalky, discolored areas on the grains known as pecky rice. Chambliss (1920) and Ito (1978) included bacteria as well as fungi. Obviously, there is need for a standard definition of "pecky."

7.1.5.1.4.1   Description of Kernel Discoloration

Douglas and Tullis (1950) stated that pecky rice could be divided into three "symptom" groups. The first type, the so-called O-type, was the insect stigmonose commonly called peck. This type was characterized by a roughly circular lesion, sometimes shrunken, and with or without discoloration. The second type, referred to as discolored or the D-type, was characterized by a general brown to black, occasionally yellow or red, discoloration of part or all of the kernel. Occasionally, small, black, sclerotial bodies might be present under the kernel coat. The third type, the L-type, was characterized by a linear black discoloration usually located at the base of the kernel but could be on the tip or side. The kernel coat generally was broken.

Helm (1954) described the damaged part of pecky rice as consisting of various discolored markings on the grain the most typical of which was known as bull's-eye peck. He noted that the damage was not visible until after the rice had been harvested and hulled. He examined samples of rice from Arkansas, Louisiana, and Texas collected during January and February, 1954, and found: (1) less damage in grain from Arkansas than from the other two states, (2) rice harvested from August 15 to September 15 in Louisiana and Texas had the most damage, (3) most damage occurred when the rice was in the milk and dough stages (approximately July 20 to September 10), and (4) rice that had not reached the milk stage by September 20 or had reached the dough stage by July 20 usually escaped major damage.

Lee and Tugwell (1980) examined samples of long-grain (Starbonnet and Lebonnett varieties) rough rice and found three primary types of damage symptoms as follows: (1) a definite brown spot on the hull with no evidence of insect damage or kernel imperfection beneath the spot, presumably caused by *Helminthosporium oryzae* Breda de Haan (= *Bipolaris oryzae*); (2) a reddish-brown hull discoloration without definite shape or boundaries; and (3) hulls that appeared normal but contained damaged kernels. They divided the latter type (3) into two subtypes: (a) those with a circular chalky area (Helm's "bull's-eye peck"?) and (b) those with black and chalky areas, usually at one end. Subtype (a) usually could be associated with evidence of hull puncture by insects, and, in fact, *Oebalus p. pugnax* previously had been associated with this type of damage; the cause of subtype (b) was unknown. Subsequently, Tugwell (1986) stated that the two most common types of rice kernel imperfections were (1) a distinctly circular, chalky, lesion 2 to 4 mm diam. and (2) a general kernel discoloration ranging from shades of black, brown, or yellow to red that might include part or all of the kernel. Although he stated that pecky rice was "a convenient catchall expression for different types of rice kernel imperfections," presumably the former type (1) was that associated with this stink bug.

7.1.5.1.4.2 Relationship Between *Oebalus p. pugnax*, Pecky Rice, and Fungi

The association between *Oebalus p. pugnax* and fungi that can cause discoloration of the kernels (i.e., pecky rice) has received much attention over the years. The feeding punctures serve as entrance points for these fungi (Douglas and Ingram 1942, Ingram 1927, Johnson et al. 1987, Odglen and Warren 1962, Robinson et al. 1980), which may enter on their own (Ingram 1927) but, more likely, are introduced by the bug (Grigarick 1984). In addition, the bugs form stylet sheaths in both the nymphal (second to fifth instars) and adult (females more than males) stages (Bowling 1979), which may facilitate entry.

Hollay et al. (1987) found the stylet sheaths generally were open, allowing fungi to enter. But, presence of the sheaths did not mean, necessarily, the stylets had successfully penetrated the hull. They found fungi in sheath openings on the interior and exterior walls of the sheaths, around the wound sites on the undersides of the hulls, and at the wound sites on the surfaces of the kernels themselves. They suggested that the fungi might use the sheaths and feeding punctures to penetrate the husks and attack kernels, thus increasing the incidence of "pecky" rice.

Marchetti and Petersen (1984) demonstrated that *Oebalus p. pugnax* feeding was a major factor in kernel discoloration. They felt that *Bipolaris oryzae* (Breda de Haan), which causes brown spot, was a secondary invader of aborted florets, possibly the primary cause of some kernel discoloration, and was just one of several microbes that colonize kernels through feeding punctures. Other fungi that have been associated with feeding of this bug, and which are capable of causing discolored areas and, thereby, "peck," include *Nigrospora oryzae* (Berkeley & Broome) (Douglas and Tullis 1950), *Curvularia lunata* (Wakker) (Douglas and Tullis 1950, Lee et al. 1986, Way 1990), *Fusarium oxysporum* Schlechtendahl, *Alternaria* spp., *Bipolaris* sp. (Lee et al. 1986, Way 1990), *Nematospora coryli, Trichoconis caudata* (Appel & Strunk), and *Cercospora oryzae* Miyake (Way 1990). Incidentally, although Way (1990) listed *Nematospora coryli* as a fungus capable of causing discolored areas, and Daugherty and Foster (1966) found it only in kernels damaged by this bug, Lee et al. (1986) were not able to create symptoms of pecky rice by inoculation with this fungus or isolate this fungus from pecky rice samples.

Lee et al. (1993) showed that discoloration in pecky rice resulted from fungi that were introduced when the stink bug was feeding, although the vector relationship was a loose one.

Attack by *Oebalus p. pugnax*, particularly in combination with fungi, results in yield loss and reduced quality. Ryker and Douglas (1938) studied the effects of certain fungi, *O. p. pugnax*, and both together on the incidence of pecky rice under relatively controlled conditions. These results were supplemented with field observations. In the field, the amount of peck ranged from 0.4 to 7.3%, and the number of bugs ranged from 93 to 3,808/acre. They noted that the high percentage of pecky rice was correlated with high *Helminthosporium* infection rather than high insect populations. However, they also noted that the highest percent of peck occurred in a field that was both heavily infected with *Helminthosporium* and had a relatively high bug population. These results were supported by greenhouse studies. Rice inoculated with *Helminthosporium oryzae* Breda de Haan (= *Bipolaris oryzae*) only, infested with *O. p. pugnax* (4 adults/cage) only, and inoculated with several species of fungi and infested with *O. p. pugnax*, concurrently, gave 3.64, 19.56, and 40.33% pecky rice, respectively.

7.1.5.1.5   Effects of Oebalus p. pugnax on Yield and Quality

Douglas and Tullis (1950) conducted caged studies of the damage to rice caused by feeding of this bug. They found the percentage of pecky rice ranged from 5 to 76% when the infestation levels ranged from 2 to 14 bugs. Also, the average number of discolored kernels ranged from 76 in the control cages to 306 in the infested cages, and actual weight of mature grains was reduced from an average of 57 g in the control cages to 39 g in the infested cages with high numbers of bugs.

Swanson and Newsom (1962) studied the effects on yield and quality of various infestation levels of late instars and adults on five varieties of rice grown in screened cages. They found consistent and severe losses in yield when bugs averaged 230 or more per 1,000 panicles. Viability of kernels was reduced because of stink bug feeding. Even at an infestation level as low as 7 to 8 bugs/1,000 panicles, feeding reduced quality to the extent that federal price support for four of the varieties was lowered. In another study, they introduced 0, 20, 100, and 500 late instars and adults into field cages containing rice in bloom. The highest infestation level resulted in 50% reduction in total yield, increased the percentage of kernel damage, and adversely affected milling yield and grade. The grade was reduced so drastically that the grain was ineligible for government price support. Grade appeared to be affected more adversely in medium-grain than in long-grain varieties. The intermediate and low infestation levels had little effect on total yield. Finally, kernels injured prior to the early dough stage often did not develop, resulting in the severe reduction in yield, noted above, for the highest infestation level.

Rolston et al. (1966) found that peckiness was more prevalent in medium-grain than in long-grain varieties, thus agreeing with Swanson and Newsom (1962). However, their results also showed that peckiness was not significantly correlated with *Oebalus p. pugnax* infestations, suggesting that medium-grain varieties were especially susceptible to peckiness from causes other than from this stink bug.

Bowling (1963) found that infestation levels of *Oebalus p. pugnax* of 0, 1, 2, and 4 adults/ft^2 in cages resulted in slight decreases in total yield and milling yield of rice and slight increases in amounts of peck at the higher infestation levels in some instances but not in others.

Robinson et al. (1980) also studied the effects of various infestation levels of *Oebalus p. pugnax* adults (0, 1, 2, 3, 4, and 5 pairs of adults/cage); the cages were infested in the early milk stage of grain development. Results showed that at the higher infestation levels the values of percent full grains and total weight (mg) per kernel were significantly less than the controls. They suggested that percent full grains should be used when comparing damage between rice varieties because total weight per kernel would be influenced by varietal characteristics.

Way and Wallace (1986a) compared *Oebalus p. pugnax* populations and the amount of peck on "Labelle" and "Lemont" varieties. In general, more bugs were collected from Labelle than Lemont. The amount of peck was not the same in all comparisons of the two varieties; in some, there was more peck associated with Labelle than Lemont. However, they found no relationship between population size and amount of peck.

Odglen (1960) and Odglen and Warren (1962), in caged studies (3 × 3 × 5 ft/cage), using infestation levels of 0, 20, 40, and 80 adults and/or nymphs/cage at the milk and dough stages found no significant difference in yields of rough, mill, and head rice among the various infestation levels and the control (no bugs). They felt (1962) that microorganisms carried mechanically by the stink bugs could have affected the milling quality of the rice had they been present. Interestingly, their results (1960, 1962) showed various percentages of peck (0.00 [= control] to 0.17) [= 80 adults and nymphs] at these infestation levels. However, we are not sure if they compared these particular values statistically, so the significance of these percentages is unknown.

Fryar et al. (1986a, b) noted there are two sources of economic loss from pecky rice: (1) price loss and (2) field loss. Price loss is affected directly by an increase in the percentage of peck, which results in a larger discount (see below). However, price is affected indirectly through a decrease in yield because of an increase in the percentage of broken heads. They also noted (1986a) that field loss occurs because many pecky grains weigh substantially less because they are not fully developed or are damaged by fungi. Because they weigh less, they often are not harvested by the combine.

As noted above, Swanson and Newsom (1962) reported that stink bugs can reduce the quality of the grain to such an extent that federal price support is lowered or the grain is not eligible for support. Lee and Tugwell (1980) noted that pecky kernels break during milling, reducing the yield of whole grain. Parboiling intensifies grain discoloration and gelatinizes chalky kernels, causing them to remain intact during milling. Additionally, expensive sorting is necessary to remove the discolored grains. Brorsen et al. (1984) stated that in three Texas rice belt bid–acceptance markets during 1978-1982, peck was the major quality factor discounting the value of rough rice, followed by red rice, weed seeds, and chalk. They (1988) and Grant et al. (1986b) further stated the most important quality factors determining the value of rough rice in Texas were head yield and peck.

Fryar et al. (1986a, b) determined that peck was a major problem in Arkansas. During 1983/1984, rice was reduced from U.S. #1 to U.S.#2 because the average lot of rice at bid–acceptance markets contained 1.34% peck (damaged by *Oebalus p. pugnax* punctures and other factors), 0.30% red rice (weedy type of rice), and 0.33% smut (rice damaged by smut fungus) in brown rice samples. They noted (1986a) that for a brown rice sample to qualify as U.S. #1 or #2, it should contain no more than 1 or 2%, respectively, of peck, red rice, and smut combined. In the present case, the combined percent was 1.97%, which barely qualified as U.S. #2. Had the peck been in the 0.30 to 0.33% range, as were red rice and smut, then the typical lot would have been less than 1% combined for the three components and, therefore, would have been graded as U.S. #1.

Speckback, characterized by various lesions that superficially resemble peck, is an imperfection that occurs on the dorsal surface of rice kernels (Bernhardt et al. 1986, 1987; Cogburn and Way 1991). However, Cogburn and Way (1991) found no evidence from their research that speckback was caused by or associated with insects.

7.1.5.2   Sorghum

7.1.5.2.1   General Information

Oebalus p. pugnax is an important pest of sorghum. Dahms (1942), in one of the earliest reports of this bug attacking sorghum, noted that it caused serious injury in Oklahoma in 1941. He first noticed the bugs (adults) on August 7 but felt they might have been present a few days earlier. The bugs were abundant until approximately August 23 and then began leaving. By September 1, very few were present. He noted that those varieties of sorghum blooming at the time of infestation were damaged most heavily. He also noted that late-maturing varieties were damaged most seriously, and that the varieties showed some differences in the extent of injury and preference by the bugs.

As with rice, both adults and nymphs can cause damage; they are attracted to the developing heads, moving from alternate hosts onto the sorghum (Fuchs et al. 1988, Hall et al. 1983). Hall and Teetes (1981) conducted a survey of plants in Texas near and in fields of sorghum from April through August. The bugs were collected only from grasses but reproduced on several host species. These hosts included, among others, oats, dallis grass, and vasey grass early in the season; and johnson grass and sorghum later in the season. The authors noted that no bugs were found in sorghum prior to flowering, the grain-developmental stage that much of the sorghum reached in early June. The bugs were collected during June from sorghum at the flowering and milk stages, although bug densities per plant generally were low during the latter half of the month. During July, numbers of bugs per field and the number of infested fields increased. Based on their survey, they considered oats and johnson grass to be important sources for the bugs during sorghum grain development because they often found large fields of these plants near sorghum, and the plants frequently were infested with the bugs. Also, johnson grass often invaded fields of sorghum.

7.1.5.2.2 Damage

Oebalus p. pugnax is attracted to sorghum, primarily during seed development. The seed developmental stages are categorized similarly to those of rice. Grain development takes approximately 36 days before reaching maturity, passing through the flowering, milk, soft-dough, and hard-dough stages, which take approximately 8, 8, 10, and 10 days, respectively (Hall and Teetes 1982a, b, c; Hall et al. 1983).

Severity of the damage to grain caused by the feeding of these bugs depends on the developmental stage of the grain when attacked, the number of bugs/panicle, and duration of infestation (Cronholm et al. 1998; Fuchs et al. 1988; Hall and Teetes 1980, 1982a, b, c; Hall et al. 1983). As with rice, the bugs cause more damage when the grain is attacked earlier in development; grain that has reached the hard-dough stage is less subject to damage (Fuchs et al. 1988; Hall and Teetes 1980, 1982a, c; Hall et al. 1983). The bugs cause the most damage when they begin their attack early during grain development and persist to grain maturity (Hall and Teetes 1980, 1982a, c; Hall et al. 1983). Attack results in reduced grain weight, quality, seed germination (Fuchs et al. 1988; Hall and Teetes 1980, 1982a; Hall et al. 1983), and grain size (Hall and Teetes 1982a, Hall et al. 1983). Extensive feeding on developing seeds usually results in undeveloped seeds that are smaller, lighter in weight, and softer than undamaged seeds. These seeds reduce bushel weight and may be lost during harvesting (Hall et al. 1983).

Although *Oebalus p. pugnax* feeds primarily on the developing seeds, it will attack other parts of the plant (Hall and Teetes 1980, 1982a; Hall et al. 1983). Feeding on the stem and rachis branches may reduce seed yield indirectly (Hall et al. 1983).

Effects of *Oebalus p. pugnax* feeding on yield of panicles at various infestation levels (0 to 16 bugs/panicle) and at various developmental stages of the grain from flowering to maturity have been investigated (Hall and Teetes 1980, 1982a, c; Hall et al. 1983). Infestation during flowering results in reduction of the number of seeds produced per panicle, whereas infestation during later grain development causes reductions in weight and size of seeds. Damage increases as infestation densities increase and is more apparent when infestations begin during the flowering or milk stage, less so during the soft-dough stage, and little, if at all, during the hard-dough stage (Figure 7.4).

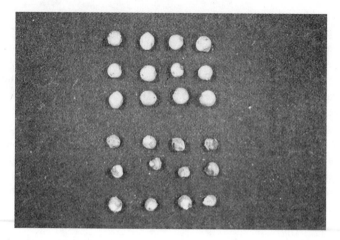

FIGURE 7.4 Undamaged sorghum seeds (top 3 rows), and damaged seeds (bottom 3 rows) resulting from stink bug feeding. (Courtesy of G.L. Teetes, Texas A&M University.)

7.1.5.3 Wheat

7.1.5.3.1 Damage

Oebalus p. pugnax has been reported from wheat since, at least, the beginning of the twentieth century (e.g., Forbes 1905, Lugger 1900), but its economic importance has not been appreciated until more recently. Viator and Smith (1980) studied the effects of its feeding on the developing heads at various infestation levels (0, 1, 3, and 6 sexed pairs of adults/15 caged heads); the insects were caged at heading and remained until the plants were harvested. The level of feeding activity was determined by the number of feeding sheaths observed. The association between the stink bug feeding sites, subsequent germination, and physical damage was determined by calculating correlation coefficients. They found that seed quality, as measured by germination and damage, was affected adversely by the feeding of the bugs.

Viator et al. (1982, 1983) followed the above experiment with a more detailed study, using similar infestation levels (0, 1, 3, and 6 sexed pairs of adults/20 caged heads) but, specifically, at the milk and soft-dough stages of grain development to examine feeding effects on kernel yield and quality. At each infestation level, they found a decrease in germination percentage, kernel weight, and baking quality as the number of stylet sheaths increased. When kernels were infested at the milk stage with as few as 1 pair of bugs, there were significant reductions in germination, kernel weight, and kernel texture. In contrast, kernels infested with 1 pair of bugs at the soft-dough stage did not show a significant reduction in these same categories. As it is rare for infestation levels to exceed 2 adults/20 spikes in the field, control measures are warranted only for wheat infested at the milk stage. Viator et al. (1983) noted in their study that the bugs produced approximately twice as many sheaths on the heads in the milk stage than in the soft-dough stage. They also noted that because the damage caused by *Oebalus pugnax* and *Nezara viridula* is virtually indistinguishable, the two taxa should be grouped together when developing control recommendations.

7.1.5.4 Soybean

7.1.5.4.1 Damage

Oebalus p. pugnax has been reported from soybean fields in several instances (e.g., Drees and Rice 1990; Miner 1961, 1966a, b; McPherson et al. 1993) but, although present, it probably was feeding on grasses within those fields and not on soybeans (Miner 1961, 1966a, b). Thus, it is not considered a pest of this crop (Miner 1961, 1966a, b).

7.2 *OEBALUS YPSILONGRISEUS* (DE GEER)

7.2.1 INTRODUCTION

The first record of this stink bug in the United States was based on two adults and one nymph collected September 20, 1983, from grass in Homestead, Dade County, FL. These specimens were reported by Mead (1983), who identified them as *Oebalus grisescens* (Sailer). However, Del Vecchio et al. (1994) now have shown that *O. ypsilongriseus* is a dimorphic species, that the two forms are the result of photoperiodic influence, and that *O. ypsilongriseus* is the long-day form and *O. grisescens* is the short-day (overwintering) form. Therefore, *O. grisescens* is not a valid species, and its name is a junior synonym of *O. ypsilongriseus*.

    Oebalus ypsilongriseus occurs throughout much of Latin America and is associated with rice (Pantoja et al. 1995). It has been reported as a pest of this crop in Colombia (Pantoja 1990) and Brazil (Del Vecchio and Grazia 1992a). In October, 1994, it was discovered in abundant numbers in rice fields at the University of Florida Everglades Research and Education Center at Belle Glade, FL, the first report of this species on this crop in the United States; it now is widespread in Florida rice fields (Cherry et al. 1998).

7.2.2 LABORATORY INVESTIGATIONS

This species has been reared from egg to adult (Del Vecchio and Grazia 1993a) and its fecundity investigated (Del Vecchio and Grazia 1992a).

7.2.3 DESCRIPTIONS OF IMMATURE STAGES

The eggs (including embryonic development) (Del Vecchio and Grazia 1992b) and first to fifth instars (Del Vecchio and Grazia 1993b) have been described.

7.2.4 ECONOMIC IMPORTANCE

7.2.4.1 Rice

7.2.4.1.1 General Information

Cherry et al. (1998) sampled eight commercial fields in the Florida Everglades Agricultural Area during 1995 and 1996 with sweep nets; fields averaged 16 ha. Sampling was conducted from 6 weeks after planting through harvest. Of the total number of stink bugs collected (n = 14,400 adults, 2,101 nymphs), *Oebalus p. pugnax* was the most abundant at 88.7% (n = 12,878 adults, 1,765 nymphs). However, *O. ypsilongriseus* represented 10.4% (n = 1,390 adults, 328 nymphs). These data clearly show that *O. ypsilongriseus* is well established in commercial rice fields in southern Florida. Also, the monthly mean number of this stink bug increased consistently during the season from May to November. Not surprisingly, *O. p. pugnax* generally outnumbered *O. ypsilongriseus* during this time. But, numbers of *O. ypsilongriseus* generally increased relative to those of *O. p. pugnax* as the season progressed until by November, *O. ypsilongriseus* actually was more abundant than *O. p. pugnax*.

7.2.4.1.2 Damage

Although no information is available on damage to rice in North America, this stink bug has the potential to attain economic pest status and needs to be monitored closely.

8 *Chlorochroa*

CONTENTS

8.1 TAXONOMIC PROBLEMS

This genus contains several closely related species in North America, many of which were confused taxonomically until the revisionary studies of Thomas (1983) and Buxton et al. (1983). Thomas (1983) divided the genus into two subgenera, *Chlorochroa* Stål and *Rhytidolomia* Stål. Within the subgenus *Chlorochroa*, he artificially divided the species into two groups, the commonly encountered Sayi group and the rare Opuntiae group. Those species of economic importance are within the Sayi group, namely, *C. C. granulosa*, *C. C. ligata*, *C. C. sayi*, and *C. C. uhleri*.

Buxton et al. (1983) revised the species of the Sayi group and included information on the geographic range of each species. The four species listed above occur in the western half of North America and have greatly overlapping ranges. *Chlorochroa granulosa* occurs from Colorado west to Nevada and north to western Montana, Idaho, Washington, and Alberta, Canada. *Chlorochroa ligata* ranges from southwestern Arkansas and Texas west to California and north to South Dakota, Montana, Idaho, Washington, and Alberta and British Columbia, Canada. *Chlorochroa sayi* occurs from western Texas west to California and north to Wyoming, Idaho, and Oregon; Froeschner (1988) added Washington. *Chlorochroa uhleri* ranges from New Mexico west to California and north to North Dakota, Montana, Idaho, Washington, and Alberta and British Columbia, Canada; Froeschner (1988) added Kansas. *C. ligata* also occurs in northern Mexico, and *C. uhleri* in Baja California. Scudder and Thomas (1987) discussed the species of *Chlorochroa* in Canada.

Much biological information has been published on several of the species, particularly on their economic importance. Many of the studies supposedly dealt with *Chlorochroa sayi* (e.g., Caffrey and Barber 1919, Patton and Mail 1935). However, because most of these studies were published prior to the revisionary works of Thomas (1983) and Buxton et al. (1983), and because the ranges of the western species greatly overlap, it is impossible to determine with any certainty which species actually were investigated. The exceptions to this are studies conducted after the revisionary studies or those conducted on *Chlorochroa ligata* in southern Texas where this species is both distinctive in appearance and does not overlap with other species of *Chlorochroa* except in the southwestern part of the state; here it occurs with *C. sayi*, but the two species can be separated easily in this geographic location (see below).

The embolium is parallel in *Chlorochroa sayi* and distinctly wider apically in *Chlorochroa ligata*. The body color of *C. sayi* is usually green, flecked with white dorsally; there are three large callosities at the base of the scutellum, one in each angle and one medially. The body color of *C. ligata* is gray or black in more southern individuals and olive-green in more northern individuals (Buxton et al. 1983). In southwestern Texas, where both species occur, only the dark gray or black form of *C. ligata* occurs (Don Thomas, personal communication); therefore, the two species are easily separated in Texas. Information purported to pertain to *C. ligata* based on studies conducted in southern Texas and adjacent Mexico probably is valid.

Based on the above, for the purposes of this work, the biological information for species of the Sayi group will be divided into two sections, *Chlorochroa ligata* and *Chlorochroa* spp. (including *C. sayi*, *C. granulosus*, and *C. uhleri*); *C. ligata* will include information only from studies conducted in southern Texas and adjacent Mexico or published after 1983. Although this is not as helpful as it would be if information could be tied to individual species, it still will serve to pull this information together and indicate where studies need to be repeated and linked to the appropriate species.

8.2 *CHLOROCHROA LIGATA* (SAY)

8.2.1 INTRODUCTION

Chlorochroa ligata, the conchuela, occurs from southwestern Arkansas, Texas, and northern Mexico west and northwest to California, and north to South Dakota, Montana, Idaho, Washington, and Alberta and British Columbia, Canada (Buxton et al. 1983).

Chlorochroa ligata attacks several cultivated crops including cotton (Morrill 1905, 1907, 1910), sorghum (Fuchs et al. 1988; Hall and Teetes 1980, 1982b, c; Hall et al. 1983; Morrill 1910), beans (Morrill 1907), alfalfa, milo maize, peaches, grapes, peas, tomatoes, pepper (Figure 8.1), squash, and corn (Morrill 1907, 1910). Uncultivated host plants include mesquite, upon which it is common (Morrill 1905, 1907, 1910), blackberry (*Rubus laciniatus* Willdenow) (Fish and Alcock 1973), raspberry, currant (*Ribes cereum* Douglas), *Pinus contorta* Douglas, *Pinus ponderosa* Douglas, Scots´ Pine (Scudder and Thomas 1987), yucca (Morrill 1907, 1910), white horse-nettle, *Ribes* sp., sage (Morrill 1910), *Arctostaphylos pungens* Kunth, *Cassia leptocarpa* Bentham, *Ephedra aspera* Engelmann, *Lycium pallidum* Miers, *Melilotus indica* (L.), *Prosopis velutina* Wooton, *Rhus trilobata* Torrey & A. Gray (Jones 1993), mallow or cheeseweed (*Malva parviflora* L.), radish (*Raphanus sativus* L.), black mustard [*Brassica nigra* (L.)], common mustard [*Brassica kaber* (DC)] (Ehler 2000), juniper, and *Opuntia* (Don Thomas, personal communication). Interestingly, although Jones (1993) reported that this species was rare on mesquite in southeastern Arizona, Thomas (personal communication) noted it was common on this plant in Texas, thus confirming Morrill's (1905, 1907, 1910) reports.

8.2.2 LIFE HISTORY

Morrill (1905, 1907, 1910) reported on the life history of *Chlorochroa ligata*, based on personal observations and on reports from growers. Combining the information in these publications, which included observations from both Texas and northern Mexico, gives some indication of the life history of this bug, but, admittedly, much additional work needs to be done. He stressed the importance of mesquite and other leguminous plants as wild hosts.

The bugs apparently overwinter as adults (Morrill 1905, 1907, 1910), which emerge during the spring and begin feeding and, apparently, reproducing on wild hosts (Morrill 1905, 1907). Although they feed on several wild hosts, they seem to prefer legumes, particularly mesquite. They will feed on leaves and stems but prefer

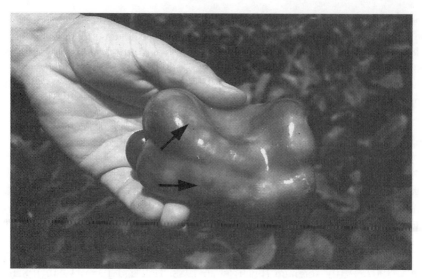

FIGURE 8.1 Irregularly shaped bell pepper resulting from stink bug damage (arrows point to feeding punctures and resulting discoloration). (Courtesy of D.B. Adams, University of Georgia.)

the beans. As these hosts mature and dry up, the bugs migrate to cultivated crops, feeding on fruits and seeds. Copulating pairs have been observed in July and early September, eggs in August, and nymphs from August into October. By November, only adults are present. These observations suggest that the species is bivoltine (recall there may be an early generation on wild hosts), but this certainly is in need of confirmation.

8.2.2.1 Mating Behavior

Fish and Alcock (1973) studied precopulatory and copulatory behavior in this species. During precopulation, the male antennates the female and attempts to lift the tip of her abdomen with his head with short pushes; this causes his body to jerk rapidly. Eventually, the male moves behind the female while vigorously antennating the venter of her abdomen. Often, the male's aedeagus is partially extruded. A receptive female lifts the tip of her abdomen, and the male then rotates, lifts his abdomen, backs into the female, and inserts his aedeagus into her genital opening.

8.2.3 LABORATORY INVESTIGATIONS

Chlorochroa ligata has been reared under confined conditions (Ehler 2000; Morrill 1905, 1910).

8.2.4 DESCRIPTIONS OF IMMATURE STAGES

The eggs and first to fifth instars have been described (Morrill 1905, 1910).

8.2.5 ECONOMIC IMPORTANCE

8.2.5.1 Sorghum

8.2.5.1.1 Damage

Chlorochroa ligata feeds on the stems, branches, and glumes of sorghum panicles but prefers seeds. Both the density of the bugs and the stage of seed development are important in the impact of feeding by these bugs (Cronholm et al. 1998; Fuchs et al. 1988; Hall and Teetes 1980, 1982b, c; Hall et al. 1983).

Sorghum grain development passes through four stages before reaching maturity: (1) anthesis or flowering (approximately 8 days), (2) milk stage (approximately 8 days), (3) soft-dough stage (approximately 10 days), and (4) hard-dough stage (approximately 10 days) (Hall and Teetes 1982a, b, c; Hall et al. 1983).

Hall and Teetes (1980, 1982b, c; Hall et al. 1983) studied the effects of feeding by these bugs during different stages of grain development and for extended periods of time. Damage was assessed by infesting panicles with various densities of adults during the milk, soft-dough, and hard-dough stages and during extended periods from the milk stage to maturity, soft-dough stage to maturity, and hard-dough stage to maturity.

The combined results from Hall and Teetes' work showed that nonseed feeding generally increased as the infestation level increased and as seed development progressed. This indicates that the bugs were less attracted to maturing seeds. Also, feeding on the panicles reduced yield and germination of seeds during the milk, soft-dough, and hard-dough stages but was most significant during early development of the seeds. At 8 bugs/panicle, the yield was reduced significantly when the infestation occurred during the milk stage but not during the soft- or hard-dough stages. Weight per 1,000 seeds generally decreased as the infestation level increased during the milk and soft-dough stages up to 8 bugs/panicle but not during the hard-dough stage, indicating the bugs had less effect on grain as the seeds developed. Reductions in germination generally increased as bug density increased. Finally, comparisons of similar densities of bugs from the milk, soft-dough, and hard-dough stages to maturity indicated that infestations from the milk stage caused the most damage to seeds.

8.2.5.2 Other Crops (Cotton, Alfalfa, and Fruit)

8.2.5.2.1 General Information

Morrill (1905, 1910) reported that *Chlorochroa ligata* was common on cotton in northern Mexico in July, 1903, peaking on July 20, and disappearing by early August. In 1904, the pattern was similar although it began a little later in the year; the bugs were scarce by the end of August. He also noted (1905) that copulating pairs were observed in early September but no nymphs and only 1 egg cluster were found around the same time; apparently, copulating pairs had been observed 5 to 6 weeks earlier in the season. The bugs fed on all parts of the plants but preferred bolls to the stems and leaves, specifically the immature seeds. Morrill (1910) included a few additional notes taken during 1905 on this bug's life history on cotton in Mexico.

Morrill (1910) reported that in 1905 the life history of this stink bug on cotton at Barstow, TX, was similar to that in Mexico. The bug was abundant in mid-July, but by mid-August its numbers had decreased dramatically. By mid-September, adult numbers had decreased slightly but nymphs, comparatively, were even scarcer. By mid-October, fourth and fifth instars were more abundant and although adults were slightly more common, most were soft, indicating they had molted recently. By mid-November, no nymphs were found and the 6 adults that were found had hardened.

8.2.5.2.2   Damage

Morrill (1907) reported that in 1904 *Chlorochroa ligata* was noted attacking alfalfa in western Texas in sufficient numbers to cause economic loss in seed crop yield. It caused similar damage in 1905. He also reported that other crops attacked seemed to have resulted from the cutting of alfalfa. In 1905, in Barstow, TX, coincident with the cutting of alfalfa, the bugs appeared on peaches shortly after July 10 as the fruits were beginning to ripen. By July 20, the peaches were heavily infested, with 10 to 15 bugs commonly observed on a single peach. These peaches became shrunken in spots, felt spongy, and fell to the ground. On July 17, 1905, in Mexico, adults and fourth and fifth instars were found in a grape vineyard with vines laden with fruit. The grape clusters were heavily infested, with a maximum of 25 bugs found on a single cluster. The bugs had migrated from an adjacent field of alfalfa.

8.3   *CHLOROCHROA* SPP.

8.3.1   INTRODUCTION

Species within the Sayi group, including *Chlorochroa ligata* (see earlier discussion), have been reported from numerous cultivated crops and native hosts. However, because of the past taxonomic problems in this genus, only a few of these plants can be associated confidently with the appropriate species.

Jones (1993) reported *Chlorochroa sayi* from *Atriplex canescens* (Pursh) (adults), *Chenopodium* sp. (adult), *Melilotus indica* (L.) (adults), and *Rorippa island-ica* (Oeder) (adults); and *Chlorochroa uhleri* from *Atriplex canescens* (Pursh) (adults, eggs, nymphs), *Atriplex polycarpa* (Torrey) (adults, nymphs, and eggs), *Cynodon dactylon* (L.) [adults and eggs; eggs parasitized by *Trissolcus utahensis* (Ashmead)], *Melilotus indica* (adults), *Poa annua* L. (adults and nymphs), and *Rorippa islandica* (adults). He further noted that Barber (1989) studied the seasonal history and host plants of *C. uhleri* (as *C. sayi*) in relation to its pest status on jojoba [*Simmondsia chinensis* (Link)] in southwestern Arizona.

Chlorochroa sayi also has been reported from tomato in California (e.g., Anonymous 1998).

Don Thomas (personal communication) has found *Chlorochroa uhleri* common on Russian thistle, *Sisymbrium*, and *Descurainia*; and *Chlorochroa sayi* common on Russian thistle, *Grayia*, *Ephedra*, *Atriplex*, and *Sisymbrium*. He also identified specimens collected in central California from tomato. He noted that *Chlorochroa gran-ulosa* often is misidentified as *C. sayi* or *C. uhleri*, and that the reports of Caffrey and Barber (1919) and Patton and Mail (1935) of *C. sayi* attacking grains probably should be attributed to *C. granulosa*. He has seen mass flights of *C. sayi* in the Mojave desert. Also, he encountered a mass flight of *C. uhleri* swarming around lights at night in large numbers at Niland, CA.

Other host plants for *Chlorochroa uhleri* include wheat, balsamroot [*Balsamorhiza sagittata* (Pursh)] (Scudder and Thomas 1987), mallow or cheeseweed (*Malva parviflora* L.), radish (*Raphanus sativus* L.), black mustard [*Brassica nigra* (L.)], and common mustard [*Brassica kaber* (DC)] (Ehler 2000).

Finally, *Chlorochroa granulosa* has been recorded from *Dryas drummondii* Richardson, *Picea* spp, and *Juniperus horizontalis* Moench (Scudder and Thomas 1987).

There are numerous reports of host plants where there is some question of the species' identities. Examples of these include wheat (*Triticum aestivum*) (Anonymous 1945, 1979; Beirne 1972; Buxton et al. 1983; Caffrey and Barber 1919; Essig 1926; Harris et al. 1941; Gillette 1904; Jacobson 1936, 1940, 1945, 1965; Knowlton 1953; MacNay 1952; Munro and Butcher 1940; Patton and Mail 1935; Pletsch 1943; Russell 1952; Strickland 1953; Twinn 1938, 1939; Wene and Sheets 1964), barley (Anonymous 1945, Beirne 1972, Caffrey and Barber 1919, Jacobson 1936, Knowlton 1953, Russell 1952, Wene and Sheets 1964), tomato (Caffrey and Barber 1919, Hoffmann et al. 1987b, Perry 1979, Russell 1952), oats (Anonymous 1945, Beirne 1972, Caffrey and Barber 1919, Essig 1926, Gillette 1904, Knowlton 1953, Russell 1952), cotton (Butler and Werner 1960; Caffrey and Barber 1919; Cassidy and Barber 1938, 1939, 1940; Clancy 1946; Essig 1926; Race 1960; Russell 1952; Telford 1957; Toscano and Stern 1976a; Wene and Sheets 1964), alfalfa (*Medicago sativa*) (Anonymous 1945, Butler and Werner 1960, Buxton et al. 1983, Caffrey and Barber 1919, Clancy 1946, Essig 1926, Gillette 1904, Jacobson 1940, Knowlton 1953, Russell 1952, Toscano and Stern 1976a), sorghum (Anonymous 1979, Clancy 1946, Hayes 1922, Russell 1952, Wene and Sheets 1964), peas (Anonymous 1945, Caffrey and Barber 1919, Essig 1926, Gillette 1904, Jacobson 1940, Knowlton 1953, Perry 1979, Russell 1952), beans (Anonymous 1945, Caffrey and Barber 1919, Essig 1926, Gillette 1904, Jacobson 1940, Perry 1979, Russell 1952), rye (Caffrey and Barber 1919, Jacobson 1936, Knowlton 1953), winter emmer, spelt, kafir corn, feterita, Sudan grass, buckwheat, cabbage, lettuce (Caffrey and Barber 1919), milo maize (Caffrey and Barber 1919, Morrill 1912), pigweed (*Amaranthus* spp.), *Stipa* spp., sheepweed (*Gutierrezia* spp.), honeysuckle [*Lonicera involucrata* (Richardson)] (Caffrey and Barber 1919), tumble mustard (*Sisymbrium altissimum*), Mormon tea (*Ephedra* spp.), hopsage (*Grayia spinosa*), tansy mustard (*Descurainia pinnata*), strawberry (*Fragaria* spp) (Buxton et al. 1983), saltbush (*Atriplex* spp.) (Buxton et al. 1983, Russell 1952), grapes (Essig 1926), beets, squash, corn, cowpea, spinach, turnips, vetch, summer cypress (*Kochia scoparia*), alfileria (*Erodium cicutarium*), canaigre (*Rumex hymenosepalus*), carelessweed (*Amaranthus palmeri*), globe-mallow (*Sphaeralcea* spp.), *Trianthema portulacastrum*, brittlebush (*Encelia farinosa*), bur-clover (*Medicago hispida*), bur-sage (*Franseria deltoidea* Torrey [= *Ambrosia deltoidea*]), Chinese pusley (*Heliotropium curassavicum* var. *osculatum*), velvet mesquite (*Prosopis velutina*), creosote-bush (*Larrea tridentata*), curlyleaf dock (*Rumex crispus*), desert sunflower (*Geraea canescens*), fragrant bitter weed (*Actinea odorata*), iodinebush (*Allenrolfea occidentalis*), johnson grass (*Sorghum halepense*), prickly lettuce (*Lactuca serriola*), wild oats (*Avena barbata* Brotero), *Gaura parviflora*, *Tidestroma oblongifolia* (Russell 1952), mallow (*Malva parviflora* L.) (Caffrey and Barber 1919, Russell 1952), Russian thistle (scientific name not given [Jacobson 1940], *Salsola pestifer* [Russell 1952], *Salsola iberica* [Buxton et al. 1983, Caffrey and Barber 1919, Ehler 2000, Munro and Butcher 1940]), tumbleweed (*Salsola kali* L. var. *tenuifolia* Tausch) (Golden and Ricker 1968), *Sisymbrium irio*, nettleleaf goosefoot (*Chenopodium murale* L.) (Hills and McKinney 1946, Russell 1952), bitterbrush [*Purshia tridentata* (Pursh)] (Basile and Ferguson 1964, Ferguson et al. 1963), dahlias (Knowlton 1953), *Chenopodium* spp. (Caffrey and Barber 1919, Munro and Butcher 1940, Russell 1952), potatoes (Beirne 1972; Daniels 1939, 1941; Gillette 1904; Knowlton 1953; Pletsch 1943; Radcliffe et al. 1991; Twinn 1939), sugar beets (Anonymous 1945; Hills 1941, 1943; Hills and McKinney 1946; Jacobson 1940; Knowlton 1953; Wene and Sheets 1964), tumble mustard (Jacobson 1940), flax (Anonymous 1945, Beirne 1972, Munro and Butcher 1940), asparagus (Essig 1926, Perry 1979), okra (Perry 1979, Russell 1952), sunflower (Anonymous 1945, Essig 1926, Gillette 1904, Perry 1979), *Artemisia* spp. (Buxton et al. 1983), sage (Essig 1926, Gillette 1904, Perry 1979), peach (Knowlton 1953, MacNay 1958), and *Opuntia* sp. (Patton and Mail 1935).

8.3.2 LIFE HISTORY

General statements about the life cycles of species in this genus are difficult because of the apparent variation in biology among species. The most informative reports, all of which supposedly dealt with *Chlorochroa sayi*, include those of Caffrey and Barber (1919) from New Mexico, Russell (1952) and Wenes and Sheets (1964) from Arizona, Patton and Barber (1935) from Montana, Munro and Butcher (1940) from North Dakota, and Jacobson (1936, 1940) and Beirne (1972) from Canada. However, as noted earlier, the taxonomy of the species in this genus was poorly understood prior to Thomas (1983) and Buxton et al. (1983). Compounding this confusion is the wide geographic range over which these earlier studies were conducted. Not surprising, then, is the variation in life cycles and behaviors reported by these earlier authors. This was first noted by Patton and Mail (1935) upon seeing differences between their results from Montana and those of Caffrey and Barber (1919) from New Mexico. Esselbaugh (1948), referring to these same two studies, stated, "There are some rather glaring discrepancies in the two accounts of the life history, causing some doubt as to whether they actually pertain to the same species."

With these rather severe limitations in mind, we present here information on the general life history of these species, excluding information on *Chlorochroa ligata*.

Chlorochroa spp. have been reported as uni- (Beirne 1972, Gillette 1904), bi- (Anonymous 1945, Jacobson 1940, Munro and Butcher 1940), and multivoltine (three or more generations; Caffrey and Barber 1919, Russell 1952), although suggestions of as many as five generations annually (Russsell 1952) are in need of confirmation.

Adults overwinter under loose bark of trees or posts (Caffrey and Barber 1919), in and around bases of native grasses (Caffrey and Barber 1919; Jacobson 1936, 1940), and in rubbish (Anonymous 1945; Caffrey and Barber 1919; Jacobson 1936,

1940; Munro and Butcher 1940), straw (Jacobson 1940, Munro and Butcher 1940, Patton and Mail 1935), stubble (Beirne 1972, Patton and Mail 1935), dead weeds (Anonymous 1945; Beirne 1972; Caffrey and Barber 1919; Jacobson 1936, 1940; Knowlton 1953; Munro and Butcher 1940), litter (Russell 1952), Australian saltbush (Russell 1952), and rocks (Jacobson 1940). Several authors have noted the importance of Russian thistle as an overwintering site (Caffrey and Barber 1919, Jacobson 1940, Knowlton 1953, Munro and Butcher 1940).

Adults emerge from overwintering sites in the spring and begin feeding and reproducing on wild hosts (Anonymous 1945; Beirne 1972; Caffrey and Barber 1919; Jacobson 1940, 1945; Munro and Butcher 1940; Russell 1952) and volunteer grains (Anonymous 1945, Jacobson 1940). At this time, the females' ovaries contain mature eggs (Caffrey and Barber 1919), which are deposited on available vegetation (e.g., rubbish, stubble, dead weeds, leaves and branches of food plants) (Beirne 1972; Caffrey and Barber 1919; Jacobson 1936, 1940, 1945; Munro and Butcher 1940; Russell 1952), including Russian thistle (Beirne 1972). The resulting nymphs feed on the young growth of wild hosts (Anonymous 1945; Beirne 1972; Caffrey and Barber 1919; Jacobson 1936, 1940, 1945; Munro and Butcher 1940). As they reach the later instars or adults, their host plants have or almost have matured and, thus, are no longer suitable as food (Russell 1952). Thus, the bugs migrate to cultivated crops that are still maturing (Anonymous 1945; Beirne 1972; Caffrey and Barber 1919; Jacobson 1936, 1940, 1945; Munro and Butcher 1940; Russell 1952). The bugs then pass through a variable number of generations until again entering overwintering sites in late fall.

Russian thistle appears to be an important wild host (Caffrey and Barber 1919; Jacobson 1936, 1940, 1945; Munro and Butcher 1940), particularly for the younger nymphs (Caffrey and Barber 1919). It is one of the earliest plants to begin growing in the spring, serves as food until late fall, and provides excellent shelter during the summer and winter (Jacobson 1940, Munro and Butcher 1940).

Patton and Mail (1935) found several differences in the life history accounts among the species they studied in Montana and the one studied by Caffrey and Barber (1919) in New Mexico, although both studies purportedly involved *Chlorochroa sayi*. They had little success finding the bugs overwintering under and around Russian thistle, unlike Caffrey and Barber, but found them by the hundreds overwintering in stubble fields, especially under straw of harvested fields. Also, the ovaries of females did not contain mature eggs as did females in New Mexico, and the eggs were deposited on stubble in preference to Russian thistle. Finally, they found several individuals feeding on *Opuntia* sp., a host plant not mentioned by Caffrey and Barber (1919). It appears, therefore, that Patton and Mail (1935) and Caffrey and Barber (1919) studied different species.

8.3.3 LABORATORY INVESTIGATIONS

Chlorochroa spp. have been reared under confined conditions (Caffrey and Barber 1919, Ehler 2000, Russell 1952). Caffrey and Barber (1919) reported on the preoviposition period, and on adult longevity, oviposition, fecundity, color change during overwintering, and several other life history parameters.

8.3.4   Descriptions of Immature Stages

The eggs and first to fifth instars have been described (Caffrey and Barber 1919).

8.3.5   Economic Importance

8.3.5.1   Crops

8.3.5.1.1   General Comments

Chlorochroa spp. begin the year feeding and reproducing on wild hosts, but as these hosts mature and dry up, and, thus, are no longer suitable as food, the bugs (generally adults and older nymphs) move to cultivated crops, primarily grains. There is no pattern as to which crops are likely to be attacked at this or other times during the season because time of attack is dependent primarily on the stage of development of each crop rather than what crops are available. Those crops in which the seeds are developing (i.e., are heading up) are preferred. Therefore, grain crops such as wheat, barley, oats, and sorghum, and nongrain crops such as alfalfa, cotton, and sugar beets, are all vulnerable at different times of the season. For example, wheat is favored several times during the year, depending on when it is planted.

8.3.5.2   Alfalfa

8.3.5.2.1   General Information

Chlorochroa spp. attack alfalfa (Caffrey and Barber 1919, Gillette 1904, Knowlton 1953, Russell 1952) and are capable of causing severe damage to immature seeds (Knowlton 1953, Russell 1952).

Russell (1952) mentioned that two alfalfa crops are grown in Arizona. The first crop is pastured or cut for hay, and a main seed crop often is produced on the second growth; the seed crop usually is started in the latter half of April and harvested in July. A second seed crop sometimes is produced on the growth that follows the main crop and is harvested approximately mid-September. Occasionally, two or three hay crops may precede the seed crop. Usually, immature seeds are present in fields from approximately mid-June to mid-September.

As noted earlier, these bugs overwinter as adults and emerge in early spring to feed and reproduce on wild hosts. The resulting adults migrate to cultivated crops (e.g., alfalfa, grain sorghum, cotton, sugar beets). As these first plants mature, the bugs move to younger crops. Russell (1952) reported that, in Arizona, the bugs were most numerous in alfalfa during the first half of July, as the crop was maturing. Peak populations occurred on cotton during the last half of August and on grain sorghum in late September or early October. Knowlton (1953) reported finding moderate populations of adults and three-fourths-grown nymphs in alfalfa seed fields in Utah on August 22. Gillette (1904) noted a report (in the form of a letter he received) of these bugs appearing in alfalfa in Colorado approximately May 20, ovipositing, and remaining there until the first cutting of hay; then leaving hay fields and migrating first to wheat fields in early July, feeding on the heads until the wheat either matured or died; and then migrating to oat fields in August, where they fed on the developing heads.

Interestingly, Caffrey and Barber (1919) stated that they were unable to rear these bugs on alfalfa from eggs to adults. Russell (1952), however, stated that he did rear them on alfalfa with immature seeds, younger alfalfa plants, and *Sisymbrium irio* L.; he gives the impression he did not rear them on each of these plants but on a combination of the three.

8.3.5.2.2 Damage

We have not found detailed specific information on the damage to alfalfa caused by these bugs other than they feed on immature seeds. When abundant, the bugs cause the seeds to become shriveled and discolored (Knowlton 1953).

8.3.5.3   Wheat

8.3.5.3.1 Damage

Adults and nymphs attack the heads of wheat where they feed on the developing kernels (Anonymous 1945; Beirne 1972; Caffrey and Barber 1919; Gillette 1904; Jacobson 1936, 1940, 1945; Knowlton 1953; Munro and Butcher 1940). Damage often is more severe in large fields nearer the margins, reflecting the bugs' migration from adjacent fields (Jacobson 1940, 1945). The bugs migrate from field to field as the heads upon which they are feeding reach maturity and those in other fields are in the process of maturing (Beirne 1972).

Feeding by the bugs can remove most of the contents from the developing kernels, resulting in shrunken and shriveled wheat (Anonymous 1945; Beirne 1972; Caffrey and Barber 1919; Jacobson 1936, 1940, 1945; Munro and Butcher 1940) and a corresponding reduction in yield (Beirne 1972; Caffrey and Barber 1919; Jacobson 1936, 1940, 1945). Loss, in addition to reduced yield, results from grade reduction because of lower quality and a decreased germination rate (Beirne 1972, Jacobson 1945).

Damaged heads assume a dull yellowish-white color (Beirne 1972, Caffrey and Barber 1919, Jacobson 1940), in sharp contrast to the bright green of undamaged heads (Caffrey and Barber 1919). Stems usually are alive and green from the base to within 5 or 6 in. of the head but yellow and dead beyond this point; the stems often break at this point during rain or wind storms (Caffrey and Barber 1919).

Feeding on the developing kernels usually is confined to the developmental period, from emergence of the head until the seed is in the late "dough" stage, when the bugs no longer can pierce the glume and kernel (Jacobson 1940). The bugs may attack during the "boot" stage (i.e., grain head beginning to form), which can slow head development and cause complete sterility of the blossoms.

Jacobson (1965) studied the effects of feeding on wheat at various developmental stages in relation to the number of insects and the duration of feeding. Feeding during the boot stage resulted in stunted plants, and feeding just before heading resulted in reduced numbers of kernels produced. Damage after heading was related directly to the numbers of insects feeding and the duration of their feeding, but inversely related to the development of the kernels at the time of feeding. Damage resulted in lower mean weights of kernels, reduced numbers of kernels per head, and discolored kernels. When plants were attacked by one or more adults/head from just after heading and in the milk stage, the resulting kernels were extremely shrivelled and yield was reduced by more than 75%. Feeding during the dough stage caused only slight damage.

Harris et al. (1941) reported that feeding by this bug resulted not only in reduced yield/acre and grade of the wheat but apparently caused severe damage to milling and baking quality in terms of lower test weight, protein content, flour yield, and bread quality. They also noted apparent differences in susceptibility to damage among varieties of wheat.

8.3.5.4   Cotton (Barley, Sorghum)

8.3.5.4.1   General Information

Chlorochroa spp. attack cotton and are capable of causing economic damage (Cassidy and Barber 1938, 1939, 1940; Toscano and Stern 1976a; Wene and Sheets 1964). They do not enter cotton fields until the plants are fruiting (Cassidy and Barber 1938). However, it seems that cotton is not a favored crop. Toscano and Stern (1976a) noted that the bugs in California (San Joaquin Valley) migrated into cotton from seed alfalfa fields. Wene and Sheets (1964) found that the bugs in Arizona moved from weeds in mid-June as those plants were drying up and into younger cultivated crops including wheat, barley, sorghum, and sugar beets. If these plants already had matured, then the bugs migrated to young cotton. If sorghum was available, the bugs moved to sorghum and then to cotton after the sorghum matured. If cotton was adjacent to sugar beets, the cotton was not infested until the sugar beets had been harvested. Wene and Sheets (1964) reported nymphs in cotton fields, indicating the bugs could survive on this crop. Russell (1952) reported that peak populations on cotton in Arizona occurred during the last half of August and on grain sorghum in late September or early October.

8.3.5.4.2   Damage

Chlorochroa spp. feed on the fruiting parts of the cotton plants, particularly on bolls, attacking the developing seeds and surrounding tissues (Wene and Sheets 1964). Injury to small bolls causes them to become soft and yellow and results in shedding (Wene and Sheets 1964). Larger bolls do not shed but will develop warty cellular growths on the inner surfaces of the carpels (Wene and Sheets 1964). The seeds may be shrivelled (Wene and Sheets 1964), and the lint may become stained (Cassidy and Barber 1938, 1939, 1940; Wene and Sheets 1964). When the soft bolls are attacked heavily, the locks may decay because of invading fungi (Wene and Sheets 1964). Stains often are visible first on immature seeds (Cassidy and Barber 1938).

Toscano and Stern (1976a), using field cages and bug densities of 0 (check), 2, 4, and 8 adults per cotton plant (approximately 0.9 to 1.1 m tall, bolls approximately 1.9-3.8 cm) per cage found that damage from the higher densities (4 and 8 adults) was most severe compared to the check, resulting in reductions of seed-cotton weight, seed weight, and lint weight compared with the control. Also, as density of the bugs increased, the weight of the cotton lint decreased.

5.3.5.5 Sugar Beet

5.3.5.5.1 Damage

Chlorochroa spp. in Arizona have been reported to migrate in mid-June from drying weeds to sugar beets (Wene and Sheets 1964). Hills (1941, 1943) and Hills and McKinney (1946), from caged studies, reported that the bugs caused a marked reduction in germination of seed produced by plants upon which they had fed and in the number of sprouts/viable seed ball.

5.3.5.6 Potato

5.3.5.6.1 Damage

Little work has been done on the damage to potato caused by these bugs. Daniels (1939, 1941) reported that feeding by these stink bugs resulted in wilting of leaves and tips. He first (1939) attributed the damage to a disease associated with their feeding (1939) but later (1941) stated that it "may be caused by the poisonous secretion of the insect during feeding." Pletsch (1943) proved the wilting of leaves or terminals resulted from the bugs' feeding and was related to density. He caged the bugs on potatoes at densities of 0, 10, 25, 50, and 100/cage and found that plants exposed to 0 or 10 bugs showed no damage; those exposed to 25 and 50 bugs, slight damage; and those exposed to 100 bugs, extensive damage.

9 *Murgantia histrionica* (Hahn)

CONTENTS

9.1 INTRODUCTION

Murgantia histrionica, the harlequin bug, is a pest of crucifers and in the past was the subject of much research because of its potential for causing serious damage. Originally an inhabitant of Central America and Mexico, it first was reported in the United States in 1864 from Washington Co., TX (Chittenden 1908, 1920; Paddock 1915, 1918; Walsh 1866; White and Brannon 1933). Its spread from Texas was closely monitored and, as a result, much information on its biology was published in the late 1800s and early part of the twentieth century. Today, as earlier, it ranges in the continental United States from New Hampshire and New York south to Florida and west to Minnesota, South Dakota, Nebraska, and California (Froeschner 1988); however, it primarily is a southern species (e.g., Hodson and Cook 1960, Osborn 1894). It now has been introduced into Hawaii (Froeschner 1988).

    Murgantia histrionica feeds on a wide variety of plants but prefers crucifers (e.g., cabbage, collard, mustard, turnip, peppergrass, shepherd's purse, radish, bitter cress, broccoli, cauliflower, kale, kohlrabi) (see McPherson 1982). Recently, Jones (1993) reported several adults on dead (flooded) *Cleome* prob. *sonorae* Gray and on *Polanisia dodecandra* (L.), and adults and nymphs on *Sisymbrium irio* L. Radcliffe et al. (1991) reported, based on a survey of potato specialists (primarily research and extension entomologists), that this bug is considered a minor pest of potatoes.

9.2 LIFE HISTORY

Murgantia histrionica is multivoltine, with estimates of the number of complete generations per year ranging from three to eight in the South and from two to five in the North, with adults usually overwintering (see McPherson 1982). These wide ranges in reported voltinism are not surprising because of the difficulty in separating overlapping generations. Also, this stink bug can become active during winter months if temperatures are mild, with feeding, copulation, and oviposition possible (Brett and Sullivan 1974), further adding to the confusion about its life cycle. However, it is probably bivoltine in the North and bi- or trivoltine in the South with an additional generation possible under favorable conditions. Recently, Ludwig and Kok (1998a) reported that it has two generations and a partial third per year in Virginia.

English-Loeb and Collier (1987) studied nonmigratory movement of *Murgantia histrionica* on *Isomeris arborea* Nuttall in southern California. They examined the importance of age and sex of the bugs and abundance of the racemes and capsules of the host plant to movement of adult bugs within stands of the plant. They found that males left the release bushes more quickly than females but, subsequently, changed locations less frequently and at a slower rate than females. There was no difference between sexes in the distance moved between recaptures. Also, there was no effect of age on time spent on the release bush, distance moved, or frequency of movement. Finally, females remained longer on those release bushes with more racemes and capsules.

McLain (1981b) studied this bug's response to varying concentrations of *Brassica napus* L. in a 5-ha abandoned field in Georgia. He found that adults preferred to colonize and oviposit within denser clusters of host plants and felt that nymphal survival was enhanced by this response.

9.2.1 MATING BEHAVIOR

Precopulatory and copulatory behavior has been reported for this species by Lanigan and Barrows (1977). The male approaches the female from the rear or front. If from the rear, he antennates the posterior part of her abdomen before moving to her head. In either the rear or front approach, he eventually begins to antennate the female's antennae. He then moves posteriorly, antennating her side and then the posterior of her abdomen. A receptive female will raise the tip of her abdomen approximately 30° and the male, with aeadeagus extended, will turn 180°, back into her, and insert his aedeagus. At any time during precopulation, an unreceptive female will simply crawl away.

9.2.2 SCENT GLAND SECRETIONS

Aldrich et al. (1996) analyzed the metathoracic scent gland secretion of this stink bug and found it included (2E,6E)-octadienedial and (2E,6E)-octadiene-1,8 diol diacetate, two natural products that were unknown in insects previously; the former was by far the major component. They also found that when the adults were squeezed, they expelled a frothy fluid from the margins of the prothorax. This fluid contains the aglucones of glucosinolates, which probably cause it to be repulsive to at least some vertebrate predators. The fluid also contains two alkylmethoxypyrazines, 2-sec-butyl-3-methoxypyrazine and 2-isopropyl-3-methoxypyrazine, which probably serve as warning odors.

9.3 LABORATORY INVESTIGATIONS

Murgantia histrionica has been reared under confined conditions by numerous investigators (see McPherson 1982), most recently by Canerday (1965) and Streams and Pimentel (1963).

9.4 DESCRIPTIONS OF IMMATURE STAGES

The eggs and various instars have been described (see McPherson 1982). Although most investigators have reported this species has five instars, Paddock (1918) surprisingly reported six.

9.5 ECONOMIC IMPORTANCE

An examination of the literature published for this species shows a remarkable decrease in the annual number of publications over the last 40 years (i.e., almost all biological information was published prior to 1960; see McPherson 1982). This indicates that the economic importance of this insect has decreased since 1960. One suspects the reason may be the same as that once seen in the decreasing importance of stink bugs as pests of cotton (see earlier discussions of *Acrosternum* and *Euschistus*); the general use of insecticides, in this case on crucifers, probably has, in fact, eliminated this insect as a serious pest. However, with efforts to decrease pesticide usage in agriculture, there is the distinct possibility there could be a resurgence in the economic importance of this stink bug.

10 *Piezodorus guildinii* (Westwood)

CONTENTS

10.1 INTRODUCTION

Piezodorus guildinii is a Neotropical species that ranges from Argentina north to the southern United States (Panizzi and Slansky 1985a); it also occurs on St. Vincent Island from which it was originally described (Stoner 1922). It has been known in the United States since, at least, the 1960s (Genung et al. 1964). Specifically, it has been reported from South Carolina (Jones and Sullivan 1982, 1983), Florida, Georgia, and New Mexico (Froeschner 1988).

Piezodorus guildinii is a serious pest of soybean throughout South America. It particularly is important in Brazil (Panizzi 1985; Panizzi and Slansky 1985b, c; Panizzi et al. 1980) where, since the late 1970s, it has begun to replace *Nezara viridula* in some areas of the country (Kogan and Turnipseed 1987, Turnipseed and Kogan 1976). It feeds on a wide range of plants, particularly legumes. Among cultivated plants, it seems to prefer soybean, alfalfa, and beans and is capable of causing severe economic damage (Panizzi and Slansky 1985a). In the United States, it has been reported from soybean (Genung et al. 1964; Jones and Sullivan 1982, 1983; McPherson et al. 1993) but usually has not been considered economically important.

10.2 LIFE HISTORY

Piezodorus guildinii has five generations per year in Argentina and overwinters as adults in protected areas. Adults are found in alfalfa from October to June, eggs from mid-November to April, and nymphs from mid-November to mid-May. Its biology is similar on soybean in Brazil (Panizzi and Slansky 1985a).

Panizzi and Slansky (1985b) found *Piezodorus guildinii* in Florida during November and December, 1983, on three plant species, which apparently were new records. On two of these, *Indigofera hirsuta* L. and *Crotalaria lanceolata* E. Meyer, they found eggs, nymphs (first to fifth instars), and adults feeding and/or basking on the pod; when disturbed, the late instars and adults hid among the pods. On the third host, *Crotalaria brevidens* Bentham, they found only one fifth instar; it was feeding on a pod.

Typical of stink bugs, the first instars cluster on or near the egg shells and do not feed. Panizzi et al. (1980), in their study of dispersal of this bug in soybeans, found that second and third instars were strongly gregarious; fourth and fifth instars were the principal instars involved in dispersal throughout the fields.

10.2.1 NATURAL CONTROL

Panizzi and Slansky (1985d) collected more than 300 fifth instars from *Indigofera hirsuta* near a mature soybean field in north central Florida and reared them to adults in the laboratory (14L:10D photoperiod, approximately 25°C and 70% RH) on *I. hirsuta* pods. From these adults, they obtained a larva of the tachinid parasitoid *Trichopoda pennipes* from a male. Earlier, Buschman and Whitcomb (1980) had reported finding this same tachinid attacking *Piezodorus guildinii*; they also collected the tachinid *Euthera tentatrix* Loew and the scelionids *Trissolcus basalis* and *Telenomus podisi* Ashmead from this stink bug. Finally, Temerak and Whitcomb (1984) added the parasitoid *Gyron* sp. to Buschman and Whitcomb's list.

10.3 LABORATORY INVESTIGATIONS

Panizzi and Slansky (1985c) reared fourth and fifth instars of *Piezodorus guildinii* to adults on *Indigofera hirsuta* pods and fed these adults on the reproductive structures of five species of legumes (i.e., *I. hirsuta* green pods, *Crotalaria lanceolata* green pods, soybean pods and seeds, green bean pods, and peanuts) in a 14L:10D photoperiod at approximately 25°C and 70% RH. They examined the differences in age of females at first oviposition, number of egg masses, total number of eggs, and egg fertility, and in adult survival, longevity, and body weight, resulting from these food sources. They concluded from these results and earlier field observations that *I. hirsuta* and *C. lanceolata* probably were important wild hosts in the seasonal phenology of this stink bug.

10.4 DESCRIPTIONS OF IMMATURE STAGES

The eggs (Bundy and McPherson 2000a) and the first to fifth instars (Grazia et al. 1980) have been described.

10.5 ECONOMIC IMPORTANCE

10.5.1 SOYBEAN

10.5.1.1 Damage

McPherson et al. (1993) studied annual variation in seasonal abundance of several stink bugs in soybean in Georgia, including *Piezodorus guildinii*, during a 5-year period (1987 to 1991). *P. guildinii* was abundant in 1987, comprising over 13% of the stink bug complex, but, during the next 4 years, it never accounted for more than 0.8% of the complex. Thus, it apparently is not the economic threat it appeared to be in 1987.

11 *Thyanta* spp.: General Information

A few published reports are available on the occurrence, biology, and damage of *Thyanta* spp. These bugs feed on a wide variety of cultivated and uncultivated plants (McPherson 1982). Crops include cotton (Panizzi and Herzog 1984, Rice et al. 1988, Schotzko and O'Keeffe 1990a), green beans (Schotzko and O'Keeffe 1990b), soybean (McPherson et al. 1993, Panizzi and Slansky 1985e), tomato (Anonymous 1998, Hoffmann et al. 1987b), peas, lentil, winter rape, alfalfa (Schotzko and O'Keeffe 1990a, b), corn, berries, sugar beets (Rice et al. 1988), and pistachio (Michailides et al. 1987, 1988; Rice et al. 1985, 1988).

Thyanta pallidovirens (Stål) feeds on the seeds of pea and lentil (Schotzko and O'Keeffe 1990a). Its feeding damage is known as "chalky spot," which is characterized by pitted, craterlike depressions in the seed coat, often accompanied by a chalky appearance of the cotyledon. It lays more eggs and has a longer life span on peas or lentils with mature pods than on those with flowers and immature pods. Further, it reproduces more successfully on peas than lentils; females averaged 91.4 eggs on peas and only 32.2 eggs per female on lentils (Schotzko and O'Keeffe 1990a).

Thyanta pallidovirens feeds on green fruits of tomato, causing a mottled appearance that becomes yellow or remains green on ripe fruits. Feeding results in destruction of cells beneath the feeding spots, which causes the damaged tissue to turn white and spongy. Damage apparently results from the injection of a digestive enzyme during feeding and the withdrawal of cellular fluids; also, yeasts and bacteria associated with the bugs' mouthparts may cause decay when introduced into the fruits (Anonymous 1998).

Sucking insects, including species of *Thyanta* (*pallidovirens*?), *Chlorochroa*, and *Acrosternum*, are responsible for "epicarp lesion" associated with pistachio fruit; the symptoms of this damage have been described in detail (Rice et al. 1985). Feeding on immature nuts by the adults or nymphs in the spring results in epicarp lesion and nut abortion. As the shells mature and the nuts begin filling during midsummer, the bugs pierce the shells and feed on the developing kernels, resulting in kernel distortion or necrosis by harvest time (Rice et al. 1988). *T. pallidovirens* feeds more frequently on the softer areas near the base of pistachio fruit (Michailides et al. 1987, 1988). It has been reared in the laboratory (Ehler 2000). Its ovipositional rhythms, including the diel mating cycle and diel ovipositional cycle, have been studied (Schotzko and O'Keeffe 1990b). In addition, its reproductive behavior has been investigated (Wang and Millar 1997). Sexually mature males of this same species release a male-specific blend of volatiles that may function as a pheromone (Millar 1997).

Thyanta custator accerra McAtee, the redshouldered stink bug, commonly is an important pest of tomatoes (Hoffmann et al. 1987b). It has been collected from numerous host plants including camphorweed, creosote-bush, tahoka daisy, alfafilla, and gramma (Jones 1993). It has been reared in the laboratory on natural and artificial diets (Brewer and Jones 1985), and its mating behavior has been studied (Drickamer and McPherson 1992).

Thyanta c. custator (Fab.) has been reported from several hosts but, unfortunately, it has had a confused taxonomic history (McPherson 1982, Rider and Chapin 1992). Therefore, it is difficult to determine which of these host records are valid.

Panizzi and Slansky (1985e) found eggs and nymphs of *Thyanta c. custator* on pods of a wild legume, *Sesbania emerus* (Aublet). They brought several third instars to the laboratory and reared them to adults on pods of this plant. Those adults that were maintained on the pods mated and produced viable eggs as did those fed mature soybean seeds and water. Finally, nymphs were reared successfully to adults on *S. emerus* and on raw, shelled peanuts but not on mature soybean seeds.

Thyanta perditor (Fab.) does not survive on soybean pods but does well on seed heads of the composite *Bidens pilosa* L. (Panizzi and Herzog 1984).

12 Management Tactics for Stink Bugs

CONTENTS

12.1 INTRODUCTION

Managing such a diverse complex of stink bug pests is a difficult task, complicated by such factors as taxonomic diversity (several genera involved), high number of host plants attacked (see Tables 2.2, 2.3, or 2.4), different biologies including life histories (e.g., host plant selection, adaptation to different climates, restriction of certain species to particular regions), and distinct responses to various control practices.

As is apparent from the life history information for the stink bug pests discussed earlier, these insects feed on a wide variety of uncultivated and cultivated plants including grasses, legumes, vegetables, and trees. Although they can feed on vegetative parts of these host plants, they prefer the reproductive structures, being most attracted to developing seeds and fruits. Therefore, they move from plants in which reproductive structures have matured to those plants in which these structures are early in development. This movement from mature plants to younger plants continues throughout the season, with the bugs feeding, reproducing, and passing through one or more generations per year.

Stink bug life histories would be of little or no importance economically if the bugs confined their feeding activities to uncultivated plants. However, this obviously is not the case. The bugs also are attracted to and damage many cultivated plants that are major agricultural commodities in the United States (Table 2.1). Thus, the bugs move back and forth between wild host plants and cultivated plants that are the most attractive to the bugs searching for a food source. Therefore, it is crucial to understand the entire life histories of these bugs, not just the periods of time when they are associated with agricultural crops.

Many crops and alternate host plants escape injury or sustain only minor levels of stink bug injury. Although this lack of economic damage often is associated with natural enemies, including entomopathogens and various environmental factors (temperature, humidity, habitat constraints), certain production practices can aid in reducing stink bug population densities and crop injury. For example, planting an early maturing soybean variety (e.g., maturity group IV) in the southern region where groups VII and VIII soybean varieties primarily are planted can avoid the higher stink bug densities and higher incidences of stink bug-damaged kernels that occur in later-maturing varieties (McPherson 1996). The early maturing varieties are nearly ready to harvest in late August and escape the highest seasonal peak populations of *Nezara viridula*, *Acrosternum hilare*, and *Euschistus servus* that occur during this time. Also, planting soybean in mid-April rather than early June lowers the seasonal mean densities of *N. viridula* and potentially avoids economic crop damage (Schumann and Todd 1982). Finally, providing suitable host plants in the ground cover of pecan orchards increases the incidence of kernel damage in the lower limbs of the trees (Dutcher and Todd 1983). Removal of weeds and cover crops in these orchards before they produce seed reduces the amount of kernel feeding on pecan (Dutcher 1990).

All of the stink bug pests highlighted in this text begin actively feeding and mating during the spring, with populations increasing shortly thereafter. This is when many crops are being planted, and most fruit and nut trees and many wild host plants are undergoing rapid growth and maturation. Population densities of these bugs intensify during the bloom and early seed formation stages of plant development. Because the host plant ranges are so broad for these pests, blooming and seed-developing host plants are available to maintain stink bug populations throughout most of the year. The year-round availability of a vast assortment of suitable host plants creates a problem for areawide management. However, the behavioral trait of stink bugs to infest plants during the plants' reproductive stages of development can be utilized to help manage these pests in integrated pest management (IPM) programs.

Because the feeding habits of stink bugs are similar, McPherson et al. (1979a) and Jones and Sullivan (1983) have stated that the best way to assess the importance of stink bugs associated with a particular crop is to consider all species together, not individually. Management tactics that have been used against stink bug pest complexes fall into four broad categories: the bugs have (1) been attacked directly with insecticides, resistant crop varieties, and biological control agents; (2) been avoided or minimized (at least the peak population densities) with early or late-maturing plant varieties and trap crops; (3) had their life cycles disrupted by destruction of weeds and other wild hosts; and (4) not been encouraged to build up damaging populations by planting preferred hosts near each other or using these hosts as ground cover in or near agriculturally important plants. Several IPM tactics and control options that have been used to manage stink bugs will be discussed.

12.2 CONTROL PRACTICES

12.2.1 Using No Control

A producer can ignore the stink bugs, but this can have serious consequences if population densities exceed the economic injury levels (EIL). The EIL is the pest density that will result in a yield loss equal to the cost of managing the pest (Higley 1994). Not controlling stink bug populations can result in large yield and quality losses in soybean (e.g., McPherson et al. 1993), cowpea (e.g., Nilakhe et al. 1981b), lima beans (e.g., Nilakhe et al. 1981c), pecan (e.g., Dutcher and Todd 1983), macadamia (e.g., Jones and Caprio 1994), rice (e.g., Harper et al. 1993), tomato (e.g., Michelbacher et al. 1952), and many other crops already covered in this review. This can be a disastrous economical decision when stink bugs exceed the EIL because it allows the bug populations to increase in abundance, damage the crops, and then migrate to adjacent later-season crops where even more severe economic damage can occur (McPherson 1996). Finally, not controlling damaging populations of stink bugs allows large overwintering populations and increased potential for damage in numerous crops the following year.

12.2.2 Biological Control

The use of biological control as a management tactic primarily involves the conservation and utilization of natural enemies to prevent outbreaks of stink bug pests or to control economically damaging populations (Todd et al. 1994). Natural enemies of stink bugs usually are present in all the crops and on alternate host plant species and may hold the pests in check for part of or even the entire season.

Stink bugs are attacked by numerous natural enemies including parasitoids, predators, and entomopathogens. In North America, most research has concentrated on parasitoids, particularly those associated with *Nezara viridula*. Many of these studies are reviewed by Harper et al. (1983), Jones (1988), Jones et al. (1983, 1996), and Luck (1981).

Although 57 parasitoid species are recorded for *Nezara viridula* worldwide, many of these parasitoids appear to be incidental; others are rare or associated more closely with other hosts or habitats (Jones 1988). Of those that are significant, the most important are the hymenopteran egg parasitoid *Trissolcus basalis* (Scelionidae) and, in the United States, the adult dipteran parasitoid *Trichopoda pennipes* (Tachinidae), actually a complex of cryptic species or biotypes in North America (Jones 1988). *T. pennipes* is a promising biological control agent of hemipterans on pecan (Dutcher 1984). Nymphal stink bugs generally are free of significant parasitoid attack

(Jones 1988). To help reduce numbers of *N. viridula* in Hawaii, *T. basalis* (Wollaston) was released in 1962 (Davis 1964, 1966) and has been released periodically since then (Oi 1991). Because of excessive damage to macadamia nuts between 1986 and 1988, an insectary was established to mass rear and release this parasitoid (Oi 1991). Several problems encountered during this rearing program included host cannibalism, temperature and humidity control, mold, and labor limitations.

Euschistus spp. (e.g., Adair 1932, Borden et al. 1952, Buschman and Whitcomb 1980, Clancy 1946, Eger and Ables 1981, Ehler 2000, Harper et al. 1983, Hoffmann et al. 1991, McPherson 1982, McPherson et al. 1982, Orr et al. 1986, Rings and Brooks 1958, Rolston and Kendrick 1961, Russell 1952, Yeargan 1979, Zalom and Zalom 1992), *Acrosternum hilare* (Drake 1920, Eger and Ables 1981, Harper et al. 1983, McPherson 1982, McPherson et al. 1982, Orr et al. 1986, Yeargan 1979), *Chlorochroa* spp. (Clancy 1946; Caffrey and Barber 1919; Eger and Ables 1981; Hoffmann et al. 1991; Morrill 1907, 1910; Russell 1952), *Murgantia histrionica* (Hoffmann et al. 1991, Ludwig and Kok 1998a, McPherson 1982), and *Piezodorus guildinii* (Buschman and Whitcomb 1980, Panizzi and Slansky 1985d, Temerak and Whitcomb 1984) also are attacked by hymenopteran and dipteran parasitoids. *Oebalus p. pugnax* (Grigarick 1984, McPherson 1982, Odglen and Warren 1962) and *Thyanta* spp. (Buschman and Whitcomb 1980, Ehler 2000, McPherson 1982, Oetting and Yonke 1971, Zalom and Zalom 1992) are attacked by hymenopteran and dipteran parasitoids and by the fungus *Sporotrichum globuliferum* Spegazzini [= *Beauveria bassiana* (Balsamo)].

Correa-Ferreira and Moscardi (1996) studied the effects on inoculative releases of *Trissolcus basalis* in early maturing soybean, which was serving as a trap crop, on a stink bug population consisting primarily of *Nezara viridula*, *Piezodorus guildinii*, and *Euschistus heros* (Fab.); in total, 15,000 *T. basalis* adults were released per hectare in the trap crop. The stink bug population density was reduced an average of 54% in the trap crop and 58% in the main crop. The parasitoids caused a reduction and delay in the overall stink bug population peak, which was held below economic levels and resulted in better seed quality.

Egg parasitism of *Murgantia histrionica* by the hymenopterans *Trissolcus murgantiae* Ashmead and *Ooencyrtus johnsoni* Howard was 8 and 37% for all eggs examined during the seasons in Virginia in 1994 and 1995, respectively (Ludwig and Kok 1998a). *T. murgantiae* was the predominate species both years and rates of parasitism steadily rose late during each season and peaked at 21 and 78%, respectively, in early September, 1994, and late August, 1995 (Ludwig and Kok 1998a).

Several predators, primarily arthropods, have been reported to feed on stink bugs in North America (Caffrey and Barber 1919, Krispyn and Todd 1982, McPherson 1982, Turnipseed and Kogan 1983, Ragsdale et al. 1981, Stam et al. 1987). Stam et al. (1987) reported 18 insect and 6 spider species feeding on *Nezara viridula*. Ragsdale et al. (1981) assessed the egg and nymphal predation of this same stink bug using an enzyme-linked immunosorbent assay and found that two predator complexes were responsible for 91.4 and 95.4%, respectively, of the egg and nymphal predation. The red imported fire ant, *Solenopsis invicta*, is reported to be an effective predator of *N. viridula*, especially on eggs (Stam et al. 1987) and early instars (Krispyn and Todd 1982). The Argentine ant, *Linepithema humile* (Mayr), and common pillbug, *Armadillidium vulgare* (Latreille), are effective predators of stink bug eggs in California (Ehler 2000). Among vertebrates, birds are reported to feed on *Oebalus p. pugnax*, *Euschistus variolarius*, and *Acrosternum hilare* (McPherson 1982). Ehler (2000) reported that mice (probably *Mus musculus* L.) fed on eggs of *N. viridula* during his field study of predation of stink bug eggs (using laboratory egg clusters) but noted this might have been an artifact (i.e., placing egg clusters too close to ground level and, therefore, unnaturally accessible to mice).

Research on the pheromones of Heteroptera has documented that tachinid fly parasitoids exploit pheromones as host-finding kairomones (chemical communicators that are advantageous to the insect) (e.g., Aldrich 1995a). These findings have enabled researchers to capture hundreds of live tachinids in traps and to study further the chemical ecology of these parasitoids (Aldrich 1995b). The pheromone of one North American general pentatomid predator, the spined soldier bug, *Podisus maculiventris*, has been identified based on patented technology and now is available (Anonymous 1992, 1999a). Pheromone secretions of several other *Podisus* species from North and South America have been identified (Aldrich 1998), and international interest is growing in using these predators in biological control programs.

12.2.3 Cultural Control

The life histories of the stink bug pest species highlighted in this review are influenced by development and growth of the fruiting structures of the host plants infested. Therefore, pest infestations often can be managed through deliberate alterations in crop production systems. Cultural practices can be utilized to adversely affect the pest species or to enhance the effectiveness of their natural enemies. Some of the cultural controls that have been used successfully to manage stink bugs include trap cropping, location of host plants, timing of planting date and plant maturation, row spacing and tillage, irrigation, resistant varieties, and clean culture.

12.2.3.1 Trap Cropping

This tactic, which is not new (e.g., Caffrey and Barber 1919), is used to purposely lure bugs into a specific area where they can be controlled effectively and economically (Hokkanen 1991). The trap crop principle encompasses the fact that stink bugs (and many other insect species) prefer certain plant species, cultivars, or crop stage(s). Manipulating these desired trap crops in time and space so they are attractive when the pests are present is essential for the success of this cultural control practice (Hokkanen 1991). It has been utilized to suppress stink bug populations on soybeans (McPherson and Newsom 1984, Newsom and Herzog 1977, Todd and Schumann 1988), pecan (Allen et al. 1995; Smith 1996a, b, 1999), and broccoli (Ludwig and Kok 1998b), among others.

On soybean, trap cropping takes advantage of the fact that stink bugs colonize this crop during the pod-set and pod-filling stages of plant development. Thus, early maturing or early planted soybean is highly attractive to stink bugs and can be used to exploit this behavioral response by attracting large populations of the bugs into small areas containing these trap crops. The heavy concentration of ovipositing adult females and the subsequent nymphal population in these crops can be effectively and economically controlled with insecticides before the next generation of adults disperses to the surrounding fields. The key to the success of this practice is to control the nymphs before they become winged adults that disperse to adjacent fields. If insecticide controls are not timed properly, then this management strategy is useless and, in fact, will intensify the stink bug problem in the main crop. Baldwin et al. (1998) recommended that no more than 15% of the planted acreage be used as a trap crop, and that an insecticide should be used when stink bug populations reach 1 stink bug per 6 ft of row or 6 stink bugs in 100 sweeps. Boyd et al. (1997) reported that if plantings of early maturing soybean comprise a large acreage on the farm each year, then the savings associated with the trap crop will be nullified. In fact, managing stink bugs could become even more intense because of the early season buildup of stink bugs on this large acreage of early maturing soybean. Recent reports indicate increasing stink bug problems on soybeans produced in the early soybean production system (ESPS) (Baur et al. 2000). The ESPS in the southern states revolves around planting soybeans in maturity groups IV and V in mid-April, approximately 4 to 6 weeks before conventional soybeans in maturity groups VI and VII are planted. The rationale behind the ESPS is to reduce drought stress on the soybean crop (Heatherly and Bowers 1998) because the early maturing varieties have produced their seeds before the excessively dry periods in July and August. McPherson et al. (1999a, 2000) reported that the ESPS contained high populations of stink bugs during August in Georgia. However, if the bugs were not controlled, they moved to adjacent conventional soybeans after the early soybeans began to mature, increased rapidly, and peaked in September and October at much higher populations than in the ESPS. Baur et al. (2000) reported similar results for Louisiana, Mississippi, and Texas.

In pecan, cowpea [*Vigna unguiculata* (L.)] can be planted on the border of the orchard and utilized to attract large populations of stink bugs (Smith 1996a, b). The stink bugs begin colonizing the trap crop as soon as the peas begin blooming and continue to feed and reproduce in the trap crop until the plants dry up. Controlling the pests, particularly the last instars, is essential for minimizing the emigration of stink bugs out of the peas and into the pecans. Feeding damage can be reduced approximately 50% in pecan orchards where trap crops are planted and stink bug population densities are moderately high (Smith 1996a, b).

In broccoli, plantings of mustard and rape can prevent low densities of *Murgantia histrionica* from moving to the main broccoli crop, but at high densities, the bugs will infest the broccoli as well (Ludwig and Kok 1998b). Therefore, both the trap crops and the main crop should be monitored to determine if an additional control (e.g., chemical) is needed.

12.2.3.2 Location of Host Plants

This tactic involves selecting which plants are grown together or adjacent to each other. Obviously, if all plants in an area are attractive to the bugs, then the potential for economic damage is increased.

To help reduce the likelihood of serious injury to alfalfa-seed crops and, perhaps, to prevent economic damage by stink bugs, Russell (1952) recommended adjustments in the layout of crops on farms, proper timing of alfalfa-seed crops, good agronomic practices, and community cooperation. Also, because many small grains and sugar beets grown for seed matured in the spring and were harvested in May or June, most of the bugs left these crops and many invaded alfalfa fields. Here, they contributed to the buildup in the main alfalfa-seed crop during June and early July. Therefore, small grains and sugar beets should not be planted near alfalfa-seed crops.

Locating the crop in a reduced tillage system also might influence the abundance of stink bugs. No-till planted corn in wheat stubble appears to sustain more stink bug damage than corn planted in conventional seedbeds (Annan and Bergman 1988).

Woodside (1950) discussed cultural control of *Euschistus* spp. (i.e., *E. servus*, *E. variolarius*, and *E. tristigmus*) regarding their attack on peaches in Virginia. He noted that peach growers could get some relief from catfacing and dimpling caused by *Euschistus* spp. by not growing crops favored by the bugs such as alfalfa, sweet clover, red clover, soybean, and cowpea.

Preferred cultivated host plants should not be planted in or near pecan orchards (Adair 1932; Mizell and Tedders 1995; Mizell et al. 1996; Polles 1977, 1979; Stein 1985) because of the potential buildup of stink bug populations. For example, interplanting cowpea in these orchards increases the amount of pecan kernel damage (Dutcher and Todd 1983). However, substituting such crops as velvet bean results in small populations and, thus, less damage (Adair 1932). Cover crops in pecan orchards, however, can be beneficial in the buildup of certain predatory and parasitic arthropods of the yellow pecan aphid, *Monelliopsis pecanis* Bissell; the blackmargined aphid, *Monellia caryella* (Fitch); and the black pecan aphid, *Melanocallis caryaefoliae* (Davis) (Bugg and Dutcher 1989). Thus, cover crops in pecan orchards can increase the severity of some pests (stink bugs) but enhance the natural control of others, such as aphids. Careful selection of plant species to be used as cover crops might allow beneficial effects without increasing stink bug populations in the orchard. Not planting truck crops, small grains, and soybean near the orchard may provide sufficient season-long protection from stink bugs on pecan (Dutcher 1990, Stein 1985).

An appreciation of the importance of location in pecan orchards began with the pioneering work of Demaree (1922), Gill (1924), and Adair (1932) and continues to be a viable option for controlling stink bugs in these orchards (e.g., Dutcher 1990). When these early investigations were conducted, much of the problem was the result of management practices (or lack thereof) in combination with a lack of understanding of the relationship between stink bugs and damage to pecan and the ability of these bugs to feed on a wide range of wild and cultivated host plants, particularly legumes. It was common practice to plant summer cover crops, which could be used for humus, or intercrops in or near the orchards. These included leguminous plants such as soybean, cowpea, bean, and *Crotalaria* spp. and nonleguminous plants such as squash and tomato. In addition, some orchards were not "sanitized," meaning that weeds (e.g., beggerweed, thistle, may-pop, jimsonweed) in and around the orchards were not controlled (see discussions of Moznette et al. 1940, Osburn et al. 1954). Therefore, bug populations could build up on these hosts and when the hosts matured, the bugs would move to the developing nuts. Thus, it was recommended that nonpreferred plants such as velvet bean be used as cover crops (e.g., Moznette et al. 1940).

12.2.3.3 Timing of Planting Date and Plant Maturation

This tactic has shown positive results in keeping stink bugs away from crops at their most vulnerable time. Stink bug populations fluctuate throughout the season relative to the phenology of soybean development. Later-maturing varieties of soybean tend to have more damage than earlier-maturing varieties, and earlier plantings tend to be damaged more than later plantings (Miner 1966a). Peaks in abundance of *Nezara viridula*, *Acrosternum hilare*, and *Euschistus servus* occur 5 to 6 weeks after the R4 (pods forming but no seeds) growth stage in all maturity groups evaluated in Georgia (maturity groups IV to VIII). Soybean varieties that mature early (groups IV and V) are colonized first because they reach R4 first, and their stink bug populations peak in late August. The later-maturing varieties become attractive (pods and seeds forming) later in the season and tend to have higher population densities, peaking in mid- to late September, due to immigration from earlier maturing host plants (McPherson 1996). Therefore, planting early maturing varieties of some crops can be used to alter seasonal mean stink bug population levels because the plants become less attractive to the bugs before the highest population peaks occur. However, insecticide applications still may be necessary to control damaging stink bug infestations in these earlier maturing crops (Boyd et al. 1997). *N. viridula* also has lower mean populations when late-maturing soybean are planted in mid-April compared with early June (Schumann and Todd 1982).

12.2.3.4 Row Spacing and Tillage

Spacing between rows of soybeans has been shown to affect seasonal mean population levels of *Nezara viridula*. Populations tend to be lower in soybeans planted in rows 90 cm apart than in soybeans planted 45 cm apart (McPherson and Bondari 1991); however, row width apparently does not influence population levels of *Acrosternum hilare* on soybean (McPherson et al. 1988). Lam and Pedigo (1998) reported no differences between no-tillage and reduced-tillage soybeans and no differences between narrow-row (25-cm spacing) and wide-row (76-cm spacing) for mean populations of both *A. hilare* and *Euschistus* spp. over a 3-year period in Iowa. No-tillage and strip intercropping in corn also does not affect populations of *A. hilare* and *Euschistus servus* (Tonhasca and Stinner 1991).

12.2.3.5 Irrigation

Stink bug populations, primarily *Nezara viridula*, *Acrosternum hilare*, and *Euschistus servus*, are significantly higher in irrigated soybean than in nonirrigated soybean in years when rainfall deficits occur (McPherson et al. 1998a). Over all the sampling dates throughout the season (averaged across 4 years), there was 31% more stink bugs in irrigated soybean (6.3 bugs per 25 sweeps) than in nonirrigated soybean (4.8 bugs per 25 sweeps), indicating that this cultural practice (irrigation) may actually enhance stink bug populations on this crop.

12.2.3.6 Resistant Varieties

Some progress has been made in developing plants resistant to stink bugs for some crops (Boethel 1999). There are several examples involving soybean. First, the cultivars Essex (Jones and Sullivan 1978), Lee 68, Hood, and Curtis (McPherson et al. 1979a) have less stink bug-damaged seeds than other cultivars. Second, Buschman et al. (1984) reported on two stink bug populations in Mississippi, one, from Poplarville, consisting primarily of *Nezara viridula*; and the other, from Starkville, consisting primarily of *Acrosternum hilare*. The Poplarville population was highest on Bragg soybeans (maturity group VII), whereas the Starkville population was highest on Tracy (maturity group VI) and lowest on Bragg (maturity group VII). Third, McPherson et al. (1996) reported that in a seven-state regional study during 1988 to 1989, stink bug population densities, primarily *N. viridula*, peaked on soybean in late summer (Julian days 260 to 270, approximately). Populations were abundant in Louisiana, Georgia, and Florida and relatively low in the test sites in the other states. Stink bugs generally were more abundant on Braxton soybeans (maturity group VII) than the other varieties studied, whereas Crockett (group VIII) and GIR81-296 (group VII) had a moderate level of resistance to stink bug infestations. Nilakhe and Chalfant (1982) screened 20 cowpea cultivars for resistance to *N. viridula* and other insect pests. Mean stink bugs per 0.92 m of row were significantly lower on the Blue Goose and the Mixed Iron and Clay cultivars than on five of the other cultivars examined, indicating that these two cowpea cultivars are less preferred plant hosts for stink bugs. And finally, the soybean plant introductions PI 171444 (Gilman et al. 1982, Kester et al. 1984), PI 171451, PI 227687, and PI 229358 (Jones and Sullivan 1979, Turnipseed and Sullivan 1976) and cowpea plant introductions PI 293557, PI 293476, PI 353074, PI 354580, and PI 293570 (Fery and Schalk 1984, Schalk and Fery 1986) have shown stink bug resistance that could be suitable in future breeding programs. Major soybean breeding efforts are being conducted in Brazil for resistance to the stink bug complex (Boethel 1999).

Schley pecans are more susceptible to stink bug damage than Moneymaker and Stuart pecans when all three varieties are interplanted in the same orchard (Dutcher and Todd 1983). Jones and Caprio (1992) examined seven macadamia cultivars for level of damage by *Nezara viridula* and found differences among the cultivars in resistance to attack. Rice breeding lines Stg 70L1188 and Stg 70L1217, out of the 228 lines screened for resistance, demonstrated a moderate resistance to *Oebalus p. pugnax* feeding (Nilakhe 1976b). Robinson et al. (1981) also screened rice lines for resistance with mixed results. Nova fresh-market tomatoes had more damaged fruit caused by stink bugs, including *Nezara*, *Chlorochroa*, and *Thyanta* spp., than 10 other tomato varieties examined (Eigenbrode et al. 1993, see tabular information). Finally, several varieties of crucifers have shown different levels of resistance to *Murgantia histrionica* (Brett and Sullivan 1974, Sullivan and Brett 1974).

12.2.3.7 Clean Culture

Clean culture has been used with some success to reduce damage by stink bugs. For example, Russell (1952) recommended eliminating overwintering habitats and uncultivated host plants associated with alfalfa fields to help prevent harmful stink bug populations from building up and seriously damaging alfalfa-seed crops. Plowing under (Caffrey and Barber 1919) or burning (Caffrey and Barber 1919; Jacobson 1936, 1940) weeds in late autumn, during the winter, or in early spring, in the latter case before oviposition up to the time the eggs hatch (Patton and Mail 1935), and spraying native plants in waste areas of cultivated fields early in the spring (Caffrey and Barber 1919) can reduce *Chlorochroa* spp. in wheat fields. Morrill (1905, 1907) suggested destroying weeds and other trash in the fall to reduce overwintering success of *Chlorochroa ligata* but felt (1907) that this would do little good in western Texas because of a high concentration of mesquite.

Destroying wild hosts of stink bugs in and around pecan orchards during the season to eliminate potential overwintering sites may provide sufficient season-long protection to pecan (Dutcher 1989, 1990; Stein 1985). Also, orchard sanitation (i.e., clean culture) should be practiced by destroying host plants in and near pecan orchards (Dutcher 1984, 1989, 1990; Phillips et al. 1964; Polles 1977, 1979), tilling soil in the orchard two or three times during summer and early fall to destroy wild hosts (Moznette et al. 1940, Osburn et al. 1954, Phillips et al. 1964), and destroying overwintering sites, particularly along fence rows (Moznette et al. 1940).

Callahan et al. (1960) noted that tomato fields overgrown with grass invariably harbored larger populations of stink bugs than clean fields. Ground cover and grass not only encourage buildup of stink bug populations, they provide some protection from chemical sprays. Therefore, clean cultivation can be of great value because it can help the effectiveness of insecticide application. Although not mentioned, the bugs undoubtedly feed on the grasses, which facilitates buildup of their populations. Replacing exotic roadside vegetation (including wild radish and mustard) with native vegetation (primarily grasses) near commercial tomato fields presumably would reduce stink bug numbers on the tomato crop (Ehler 2000). Mowing the roadside weeds, either prior to colonization by *Euschistus conspersus* or when the population is comprised primarily of nymphs, also would help to manage this stink bug. However, this roadside vegetation also appears to be critical for the buildup of certain arthropod predators and stink bug egg parasitoids (Ehler 2000).

Scott and Madsen (1950), Borden and Madsen (1953), and Borden et al. (1952) recommended clean culture in and around pear orchards because *Euschistus conspersus* populations increase on weeds and cover crops. Thus, if these host plants are eliminated, populations also will be reduced because the bug does not reproduce on pear.

Woodside (1947) stated that *Euschistus servus* and *E. tristigmus* did not breed on peach, and the weeds apparently were more important than cover crops as a source of stink bugs on peach. Therefore, he recommended that if cultivation was not practiced, weeds should be kept mowed to reduce the reproduction of these bugs. He also noted (1950) that because damage to peaches from *A. hilare* often was highest in those trees bordering woods, removing wild hosts from the edges of the woods could be helpful in reducing the number of bugs that could migrate from those hosts to the peach trees. The more important of the wild hosts included elderberry, black locust, honey locust, linden, redbud, wild cherry, dogwood, and boxelder. He also suggested (1950) that monetary losses from catfacing caused by bugs could be reduced by removing scarred fruits during thinning before removing other fruits.

Keeping weed populations down, particularly grasses, on levees and in and around the rice fields (Ingram 1927, Naresh and Smith 1984, Webb 1920) and destroying potential overwintering sites (Douglas and Ingram 1942, Ingram 1927, Naresh and Smith 1984) can help in reducing *Oebalus p. pugnax* populations on rice.

Finally, Ludwig and Kok (1998b) suggested that destroying crop residue after harvest of broccoli, which serves as a food source for *Murgantia histrionica* before overwintering and until the next planting, could help reduce populations of this stink bug the following year.

Interestingly, clean culture is not always desirable. A weed-free clean culture in soybeans actually can increase the likelihood of *Nezara viridula*. Altieri et al. (1981) reported more of these bugs in weed-free soybeans than in soybeans left weedy. A relationship was observed between stink bug abundance and periods of weed-free maintenance. Lowest bug numbers were observed in plots where weeds remained in the field the entire season, intermediate numbers in the plots where weeds were controlled for either 2 or 4 weeks after planting, and highest numbers in plots kept weed-free for the entire season.

12.2.4 Chemical Control

Over 151 million pounds of active ingredient of insecticides were sold in the United States in 1998, at a cost of over $1.83 billion. The corn, cotton, and vegetable agroecosystems accounted for 72.6% of the total crop market (Association Survey Resources 1999). The stink bug pests of economic importance accounted for some of these costs. In Georgia alone in 1996 (the last published report), stink bugs cost producers over $5.2 million, $3.1 million, and $0.4 million in corn, cotton, and vegetables, respectively, and stink bugs were the most important economic pest in corn that year (Riley et al. 1997).

Chemical control techniques have changed dramatically during the twentieth century, both in the types and amounts of chemicals used and methods of application. For example, in 1942, Watson and Tissot gave the following control recommendation in reference to control of *Nezara viridula*: "For hand collecting use a shallow pan and pour into it an inch or so of water and on this a film of kerosene. In the early morning or on a cool rainy day when the bugs are sluggish walk along the rows and knock them into the pan. This is not as slow a process as it may seem at first. An alert boy ("active boy," Watson 1919) can collect most of the bugs from an acre of beans or cowpeas in an hour or two." Morrill (1905, 1907, 1910) recommended similar control techniques for the time (e.g., handpicking, knocking insects into a container) for controlling *Chlorochroa ligata*. Although outdated, these recommendations make interesting reading. Obviously, other stink bug control options are more effective and much less time consuming.

Insecticides are classified primarily on the basis of their chemical structure. The inorganic insecticides contain no carbon in their chemical structure, whereas the organic insecticides contain carbon and usually are produced synthetically, although some are naturally produced from plants, soil, or microorganisms. The main classifications of insecticides include organic phosphates or organphosphates, carbamates, chlorinated hydrocarbons, or organochlorines, synthetic pyrethrins, or pyrethoids, botanicals, microbials, insect growth regulators, and fumigants (Guillebeau 1998); plus oils, neonicotinoids, and pyridine azomethines.

Conventional chemical control generally has provided an effective and economical means of suppressing phytophagous stink bugs and, in fact, remains the only consistently reliable practice for control of these pests on soybean and other crops (e.g., Delaplane 1995, Funderburk et al. 1999). Numerous reports on the use and efficacy of insecticides on stink bugs have been published, the majority of which deals with control of *Nezara viridula* on soybean and other row crops, pecan, and vegetables (e.g., Anderson and Teetes 1996; Boyd et al. 1996; Callahan et al. 1960; Chalfant 1973; Chalfant and Young 1982; Clower and Hankins 1960; Fitzpatrick and Boethel 1998; Hoffmann et al. 1987b; Lingren et al. 1995; Marsolan and Rudd 1976; McPherson et al. 1979b, 1993, 1995a, b, 1998b, 1999b, c; Polles 1977, 1979; Schuster 1991, 1998; Schuster and Polston 1998a, b; Southwick et al. 1986; Taylor et al. 1998; Todd and Schumann 1988; Way and Wallace 1995; Zalom and Zalom 1992), and control of *Oebalus p. pugnax* on rice (e.g., Bowling 1956, 1962, 1967; Brook 1953; Drees 1983; Fryar et al. 1986a, b; Gifford et al. 1968b; Grigarick 1984; Hamer and Jarratt 1983; Oliver et al. 1972; Robinson et al. 1987; Sudarsono et al. 1992; Texas Agric. Ext. Serv. 1997; Tugwell 1986; Way and Wallace 1986b, 1990; Way et al. 1987) and other grains (e.g., Anderson and Teetes 1996, 1997; Anderson et al. 1998). Reports for most of the other species reviewed in this work also have been published (e.g., Borden and Madsen 1953; Borden et al. 1952; Callahan et al. 1960; Cassidy and Barber 1938, 1939, 1940; Dutcher 1987, 1990; Griffin and Khan 1998a, b; Hoffmann et al. 1987b; Hogmire and Winfield 1998; Khan and Griffin 1999; McPherson et al. 1979b, 1995b; Russell 1952; Southwick et al. 1986; Walgenbach and Palmer 1998, 1999a, b; Walker and Anderson 1933; Way and Wallace 1995; Zalom and Zalom 1992). Several of these reports are outdated but interesting from a historical perspective. The importance of correctly timing the spray schedule has been discussed (Jones and Cherry 1984).

Interestingly, Way (1990) noted that little effort has been made to develop nonchemical controls for *Oebalus p. pugnax* for several reasons, including the short period of host plant vulnerability (i.e., heading to harvest, which is approximately 30 days for most varieties), the high mobility of the bug, the low economic threshold densities, and the relatively low cost of chemical controls.

12.2.4.1 Application of Insecticides

Insecticides usually are applied by either a fixed-wing aircraft or a tractor-mounted sprayer with relatively low pressure and low volume. However, a backpack sprayer often is used in small-plot test evaluations or for localized treatments of small areas. Chemigation also can be used to control stink bug pests. For example, it has been used effectively with certain insecticides to control *Nezara viridula* on processing tomatoes (Chalfant and Young 1982), although control with the newer synthetic pyrethroids has not been effective (Schuster 1991).

Hoffmann et al. (1987b) compared the effectiveness of aerial and ground sprays in reaching stink bugs usually found low on tomato plants. They discovered that insecticidal coverage in the upper canopy was over three times better than at ground level when applied by ground equipment and approximately seven times better than at ground level when applied by aerial equipment. These results indicated that little material reached the part of the plant canopy (lower) occupied by approximately 40% of the bugs and, therefore, control with insecticides was inadequate. McPherson et al. (1979b) reported controlling stink bugs, including *Nezara viridula*, *Acrosternum hilare*, and *Euschistus servus* through aerial application of methyl parathion at rates of 0.28 and 0.56 kg AI/ha. Both rates gave effective control of all developmental stages of all three species except *A. hilare* fifth instars and *E. servus* fourth and fifth instars. This was confirmed in the laboratory, in part, where it was found that the ug/insect LD50 (topical application) was significantly higher for the fifth instars than the adults in all three species.

Chemical control of stink bug species on fruit and nut trees requires different techniques than those used for row crops and vegetables. Insecticides are applied with an airblast orchard sprayer at high pressure and high volume. Sprays often are applied on a preventive basis on a 11 to 14 day spray interval (Dutcher 1990) due to the difficulty in sampling to determine if an economic injury level or economic threshold of stink bugs is present. However, spray decision guidelines are being developed based on stink bug numbers present, damage observed, developmental stage of the crop, and previous yields under similar conditions (da Silva and Daane 1995).

12.2.4.2 When Should Insecticides be Used? Sampling and Economic Thresholds

Insecticides should be used only when necessary, using EILs or economic threshold levels and at recommended minimum effective dosage rates; this will provide acceptable control and preserve many of the natural enemies (Higley 1994, Todd et al. 1994). However, determinations of EILs and economic threshold levels require estimations of the population levels of the pest species present. Several sampling methods are available and include use of a sweep net (Figure 12.1), ground cloth (shake cloth) (Figure 12.2), beat-bucket, and, more recently, pheromone trap (Figure 12.3).

Bowling (1969) found a high correlation between visual counts of *Oebalus p. pugnax* and sweeping with a 15" (38 cm) diam. insect net. Therefore, using the formula $Y = a + bX$, he was able to estimate population size from sweep counts. This is important because it means sweep counts can be used to estimate population levels and, therefore, determine when insecticides should be applied. Using routine field monitoring (Bowling 1969, Smith et al. 1986) and economic treatment threshold levels (Drees 1983; Hamer and Jarratt 1983; Harper et al. 1993, 1994; Texas Agric. Ext. Serv. 1997) to determine if and when insecticide

FIGURE 12.1 Using a sweep net to sample plants for stink bug pests. (Courtesy of R.M. McPherson, University of Georgia.)

FIGURE 12.2 Using a ground cloth (shake cloth) to sample plants for stink bugs. (Courtesy of R.M. McPherson, University of Georgia.)

applications are necessary are efficient and economical pest management techniques for managing *O. p. pugnax* on rice (Harper et al. 1990). These insect management decision guidelines, and the economic conditions of the crop, have

FIGURE 12.3 Pheromone trap (designed by R.F. Mizell) used for monitoring stink bugs in soybeans. (Courtesy of J.K. Greene, University of Georgia.)

been incorporated into a dynamic IPM program for *O. p. pugnax* on rice in the southwestern United States (Harper et al. 1994). Incidentally, Way (1984) noted that because of increased yields from "Lemont," coupled with stable rice prices, the economic threshold for *O. p. pugnax* was expected to decrease. Harper et al. (1993, 1994) presented results of their work to identify flexible economic thresholds for this bug on rice.

Hall and Teetes (1982c) and Hall et al. (1983) determined equations $[E(Y) = bX^2]$ to estimate percentage of sorghum yield loss at different infestation levels of *Oebalus p. pugnax*, *Nezara viridula*, *Chlorochroa ligata*, and *Leptoglossus phyllopus* (leaffooted bug) and calculated economic injury levels. Cronholm et al. (1998), Hall et al. (1983), and Fuchs et al. (1988) further calculated injury levels for different control costs and crop market values. Control is based on the stage of grain development at the time infestation occurs, the number of stink bugs/panicle, the cost of controlling the bugs (insecticide and application), and the market value of the grain (yield × price). They provided tables to determine the per panicle economic injury levels for *O. p. pugnax* (bugs/panicle) at the flowering, milk, and soft-dough stages when controlling an infestation would be justified.

The above economic thresholds require estimating the number of insects per sorghum panicle. Merchant and Teetes (1992) determined the "beat-bucket" sampling technique was a simple but highly efficient method for determining the number of *Oebalus p. pugnax* individuals per panicle. The technique involves grasping the peduncle just below the panicle and vigorously shaking and striking it against the side of a bucket for 4 to 5 seconds to dislodge the insects. To test its efficiency, the panicle was covered with a clear plastic bag, excised, taken to the laboratory, and examined for missed stink bugs. They found the technique captured 100% of adults, 98% of large nymphs, and 95% of small nymphs.

Lye et al. (1988b) developed polynomial regression models for estimating injury to tomatoes by *Nezara viridula*. The percentage of fruit to be classified as U.S. grade 1, 2, or 3, at a particular bug density and feeding duration, can be estimated from these models. Not surprisingly, they found the threshold for cosmetic damage is much lower than the threshold for weight loss and fruit size reduction. Pheromone traps can be used to detect migration of stink bugs into tomato fields and to determine when to begin sampling the crop with a beat sheet or beat tray. In tomato fields that have mostly large green fruit, damage from stink bugs increases by 6% for each stink bug collected in the beating tray (16-in. cafeteria-style tray) (Anonymous 1998). The rate of damage is slightly higher when pink fruits are present. Stink bug damage does not increase rapidly because the bugs tend to concentrate in one area; thus, the producer has some time to make a treatment decision (Anonymous 1998). Insecticides are more likely to be needed in fresh market tomatoes and for tomatoes that will be used for whole peel processing; they are not recommended for tomatoes intended for paste except when conditions are favorable for the development of yeast or fungal pathogens that can be introduced by stink bug feeding (Anonymous 1998).

Pheromone traps also can be used in pecan orchards to help determine if and when controls are necessary (Mizell et al. 1996, 1997). Stink bugs can attack the pecan crop anytime proper conditions are present. Therefore, it is essential to monitor individual orchards from June to harvest for the best management decisions. Three to five pheromone traps on both a border row and in the interior of the orchard apparently will provide useful information for making these decisions (Mizell et al. 1997).

Todd (1981) indicated that control is recommended on soybean when stink bug populations exceed 1.1 bugs per row-m up to mid-podfill, then 3.3 bugs per row-m from podfill to maturity. However, when the crop is produced for seed, control is recommended when populations reach 1.1 bugs/2 row-m. Conversions for comparing the numbers of stink bugs per row-m to numbers per 25 sweeps have been reported for *Nezara viridula* (Todd and Herzog 1980), and for *Euschistus servus* and *Acrosternum hilare* combined (Deighan et al. 1985). Sane et al. (1999) recently reported on the efficiency of sweep net, ground cloth, vertical beat sheet, and absolute sampling methods for estimating stink bug (all species combined) population densities, and several other arthropod pests as well, in conventional wide-row (95-cm spacing) and close-row (17.5-cm spacing) soybeans. Their results indicate that both sweep net and ground cloth provide good estimates of stink bug densities (regression coefficients $r^2 = 0.85$ and 0.75 in conventional soybeans for sweep net and ground cloth, respectively; and $r^2 = 0.82$ and 0.71 in close-row soybeans for sweep net and ground cloth, respectively). The vertical beat sheet had much lower r^2 values (0.65 for conventional and 0.44 for close-row soybeans).

Race (1960) compared the effectiveness of sweeping to collection with a mechanical device for estimating population size of *Chlorochora* spp. in cotton.

Ellis (1984) reported on the techniques for sampling pecan pests, including stink bugs, to obtain an accurate estimate of the pest infestations. They stated that the samples should be objective, consistent, adequate in number, frequent enough, extensive enough, and practical with regard to time and effort. To obtain an accurate, unbiased orchard survey of insect pests, they gave the following guidelines:

1. Sample trees in all segments of the orchard (approximately 10% of the trees).
2. Identify and sample each major cultivar in the orchard.
3. Sample at least five compound leaves and five nut clusters per tree and carefully inspect for pests and damage.
4. Take samples randomly.
5. Increase the number of samples per tree when pest populations are near treatment levels to improve confidence in control decisions.
6. Sample each orchard at least once per week.
7. Keep good written sampling records on each orchard.
8. Be observant when traveling through the orchard so you do not overlook other problems that are not a part of the pest survey.

12.2.4.3 Negative Effects of Insecticides

Several of the standard materials used for controlling stink bugs have been removed from the marketplace, or are pending removal, due to label revision or cancellation because of environmental and human safety concerns or unjustified costs of the re-registration process (Todd et al. 1994). Alternative insecticide controls are being examined and include new-generation pyrethroid and insect growth regulators (Lingren et al. 1995, McPherson et al. 1995b, Wier et al. 1991). These newer products, at dosage rates of 0.028 kg AI/ha and lower, have increased efficacy on *Nezara viridula* control and are less detrimental to the environment and many beneficial species. The efficacy of transgenic Bt cotton on certain cotton pests, including *Acrosternum hilare*, also is being investigated (Mahaffey et al. 1994).

Unfortunately, insecticides can have a negative effect on arthropod parasitoids and predators. Sudarsono et al. (1992) studied the survival of 2 to 3- and 6 to 8-day-old immature *Telenomus podisi* parasitoids and *Oebalus p. pugnax* embryos after field application of methyl parathion and carbaryl in relation to the animals' locations in the rice canopy. *T. podisi* immatures located in the upper canopy had lower survival than those in the lower canopy, regardless of age. *O. p. pugnax* embryos in the upper canopy, compared with *T. podisi* embryos, were significantly more susceptible to the insecticides. Insecticides had no effect on immature *T. podisi* or stink bug eggs located in the lower canopy. Methyl parathion reduced survival in both species more than carbaryl. Smilanick et al. (1996) found that emergence of *Trissolcus basalis* from egg masses of *Nezara viridula* exposed to methamidophos residues on tomato foliage was not affected, but survival of the emergent adults was reduced.

Insecticides also can lead to the resurgence of other arthropod pests, primarily because of the reduction in numbers of natural enemies (Panizzi and Slansky 1985a, Shepard et al. 1977).

The cultural control methods and natural enemies discussed above suppress stink bug populations. These, in turn, help reduce insecticide use and, thereby, avoid insecticide resistance problems with many insect pests (Todd et al. 1994). However, even with extensive use, insecticide resistance is not always the cause for chemical control failure. In an interesting report by Drees and Plapp (1986), a possible case of insecticide resistance in *Oebalus p. pugnax* proved to be something quite different. Rice growers in two counties in Texas were having difficulty controlling this bug relative to growers in a third county. Responses of bugs from the first two counties to carbaryl and methyl parathion compared with the third county showed no toxicological differences in calculated LC_{50} and LC_{90} values. The authors suggested that the problem was with the size of sorghum plantings. Sorghum is an alternative host to rice for *O. p. pugnax*. They noted that in the county where control was not a problem, less than 500 acres (202 ha) of sorghum had been grown annually since 1980. In the other two counties, an additional 84,000 acres (34,000 ha) were planted in sorghum in 1981 and rice production declined by 59,000 acres (24,000 ha) between 1981 and 1983. Therefore, sorghum greatly increased relative to rice. During years when sorghum matured and was harvested as rice grain was maturing, the bugs migrated in mass to the rice. Where more sorghum was grown relative to rice, the migration pressure was more severe, regardless of the insecticide used.

Even with the negative aspects associated with chemical control, use of insecticides as a management practice remains a vital component of the Integrated Pest Management programs for many crops throughout the United States. This particularly is true in the southern region, where economically damaging infestations of *Nezara viridula* and other stink bug species are annual economic threats to several crops. Most of the insecticides labeled today have short residual activity compared with the older chlorinated hydrocarbons, thus greatly reducing the amount of insecticide residue remaining in the agroecosystem.

Changes in Environmental Protection Agency policies and the Food Quality Protection Act, patent expirations and label uses, worker protection standards, and economic/environmental concerns necessitate the constant evaluation and re-evaluation of insecticidal controls. Insecticide efficacy, residual activity, impact on non-target organisms, and environmental quality are constant concerns to research and extension personnel working on the management of stink bugs (Chyen et al. 1992, McPherson et al. 1995b, Way and Wallace 1986b). Insecticide trials are conducted annually on stink bug control on many crops, and many of these tests are reported in the annual issues of the Entomological Society of America's Arthropod Management Tests (formerly, Insecticide and Acaricide Tests).

12.3   CONCLUSIONS

Managing economically damaging infestations of stink bugs is a complex and diverse activity. It requires a knowledge of the cropping systems involved, the economics of control practices and potential damage losses, the environmental impact of various management practices, the ecological and behavioral responses associated with specific plant/pest interactions, and the impact of cultural practices on current and subsequent stink bug pest populations. All of the stink bug pests reviewed in this work are attracted to the reproductive growth stages of their host plants. Until the crops of major concern reach the desired reproductive stages, the stink bugs feed and reproduce on numerous wild (e.g., weeds, trees, berries) and other cultivated (e.g., winter grains, early-season vegetables) plants. Then they move into the major plantings of row crops, vegetables, and fruit- and nut-producing trees where extensive economic losses can occur because of pesticide control costs and quality/yield reductions.

Many crops escape major stink bug injury or sustain little damage because of various biotic (i.e., parasitoids, predators, entomopathogens) and abiotic (e.g., temperature extremes, humidity, habitat availability) factors that keep the bug populations low directly or indirectly. Crop phenology also may not be synchronized with the stink bug's life cycle. However, if an economically damaging infestation of stink bugs develops, several management options are available.

One option is for the producer/processor/consumer to tolerate the stink bug problem and accept the crop quality and yield reductions that occur. Another option is to tolerate the infestation until an EIL is reached, then treat with an insecticide to keep crop losses to a minimum.

Other practices can be applied to avoid stink bug infestation or damage in the primary crop. These include planting trap crops, planting early or late-maturing varieties that escape stink bug migrations, and planting varieties that are resistant to stink bug feeding. Finally, destroying alternate host plants in or around cultivated fields and avoiding the planting of certain ground cover crops in orchards can aid in managing these pests. All of these practices help to minimize the use of insecticides, thereby helping to protect the environment.

12.4   FUTURE OUTLOOK

Although the future pest status of stink bugs is uncertain, several emerging crop production technologies and crop protection strategies raise concerns that stink bug problems may increase in coming years in certain crops. With the successful elimination of the boll weevil as an economic pest across most of the cotton belt, the insecticide applications in cotton have been reduced dramatically and, in some instances, even eliminated. The recent registration of Bt cotton also has provided an opportunity to reduce or eliminate insecticide use directed at the bollworm/budworm pest complex. Finally, cotton IPM programs are placing more emphasis on biological control of various pests, including tobacco aphids, and, thus, fewer insecticide sprays are being applied. Stink bug populations and resultant damage are increasing on cotton throughout the Southeast (Bacheler and Mott 1995), suggesting that the numerous insecticide sprays for boll weevil, bollworm, and aphid control also had been suppressing stink bug populations.

Because there is little economic benefit to controlling corn earworm and many other pests on field corn (Sparks and Mitchell 1979), foliar insecticide applications have been reduced, especially with the recent registration of Bt corn. This reduction has provided a safe haven for early season stink bug population buildups in field corn prior to the bugs migrating to soybean, vegetables, and other late-season crops where subsequent generations rise to high densities.

The impact of herbicide-tolerant biotechnology on the dynamics of stink bugs currently is unknown. These new technologies (Roundup Ready, Liberty-Link, STS genetic insertions, and others) in corn, cotton, and soybean may have major positive or negative effects on stink bug populations by providing a relatively weed-free crop environment. Whether or not these new weed management options will increase stink bug damage to crops (by removing the direct association of alternate weed hosts in the field) or, perhaps, make crops less susceptible to stink bug colonization and population buildup is uncertain at this time.

Other biotechnological advances in crop production and crop protection may have major impacts on stink bug abundance and survival as well. These particular advances could include plant genetic alterations for specialty markets (e.g., more protein, increased lysine, decreased sugars, reduced polyunsaturates, increased oleic acid, better milling, and smaller seed size), host plant resistance to pathogens and arthropod pests other than stink bugs, drought tolerance, use of the Global Positioning System and precision farming, and other areas for which no information is currently available on how stink bug populations will be affected. Changes in production environments (e.g., new tillage systems, ultra-narrow row spacings, trickle irrigation systems, plastic mulch, and earlier- or later-maturing varieties) may also have positive or negative impacts on stink bugs and other arthropod pests.

Implementing the recent discoveries of semiochemicals into IPM programs could also provide an environmentally sound approach to stink bug management. For example, pheromones for important generalist arthropod predators could be used to attract these beneficial organisms into specific areas where damaging pest populations are present. Conversely, pheromones could be used to lure predators away from pest infestations prior to insecticidal applications, thereby conserving these predators for later use against resurging pest populations (Jeff Aldrich, personal communication). Adding pheromones or prey-associated kairomones (semiochemicals that are advantageous to an insect for host finding, feeding, and oviposition) into artificial diets and oviposition sites might improve the efficiency of predator mass-propagation (Aldrich, personal communication). Attractants for phytophagous stink bugs would provide an effective method for monitoring infestations in many crops (now in use for tomatoes in California [Anonymous 1998]) and could help to concentrate these bugs into small areas, such as into trap crops or early maturing varieties. Research in California and Florida indicates that trapping immigrant adult phytophagous bugs will suppress infestations. A trapping scheme is being developed to harvest predatory stink bugs in early spring for augmenting biological control and to capture and destroy pest stink bugs later in the season using the same traps with different pheromones (Aldrich, personal communication). In the United States, attractant pheromones or suspected pheromones have been identified for *Nezara*, *Euschistus*, *Acrosternum*, *Murgantia*, and *Thyanta* species (Aldrich 1995a, Aldrich et al. 1996, Millar 1997). Monitoring kits based on patented technology for *Euschistus* spp. are available (Anonymous 1999b, c).

Release of low-level radiation to sterilize stink bugs also might be used in the future if this technology proves successful as a tactic to suppress stink bug population densities in an areawide pest management program (Dyby and Sailer 1999).

Piezodorus guildinii populations now are present throughout Florida and southern Georgia. This stink bug is a serious pest on soybean throughout Brazil and, potentially, could attain key pest status on soybean throughout the southern United States in the future.

Oebalus ypsilongriseus populations now are widespread in Florida rice fields. This stink bug occurs throughout much of Latin America and has been reported as a pest of rice in Colombia and Brazil. Therefore, it potentially could attain key pest status on rice produced in North America.

Finally, other stink bug species, which already are present in low numbers in the United States or which may be introduced in the future through foreign trade, could also adapt to the growing conditions and crops being produced in North America and become economic pests. For example, there are several species that are important pests of citrus in Australia, Asia, and probably other continents as well, and it would not be surprising to find they could adapt well to the citrus-growing areas of North America.

Although much research has been published on the Pentatomidae, and much of this work is cited throughout this book, additional research is needed in the areas outlined in the future outlook section. The entomologists who have published their findings on stink bug identification, life history, behavior, population dynamics, and damage are to be applauded for their efforts during the twentieth century, but many exciting challenges and opportunities await those who pursue the study of these insects in the twenty-first century.

Appendix

Annual estimated losses in Georgia soybean from stink bug damage and control costs, 1971 to 1998. Asterisks identify those years when stink bugs were the most damaging soybean insect pest. (Data from annual Georgia Agricultural Experiment Station Special Publications on statewide insect losses.)

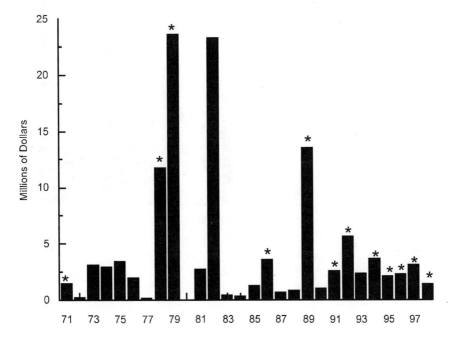

Figure A.1

Annual estimated losses in Georgia corn from stink bug damage and control costs, 1971 to 1997. Asterisks identify those years when stink bugs were the most damaging insect pest on the crop; n/a, data not available. (Data from annual Georgia Agricultural Experiment Station Publications on statewide insect losses.)

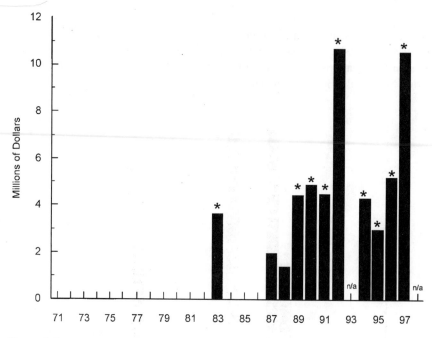

Figure A.2

Annual estimated losses in Georgia cotton from stink bug damage and control costs, 1971 to 1998. Stink bugs have never been the most damaging insect pest on this crop. Bt cotton was available commercially to growers beginning in 1996. (Data from annual Georgia Agricultural Experiment Station Publications on statewide insect losses.)

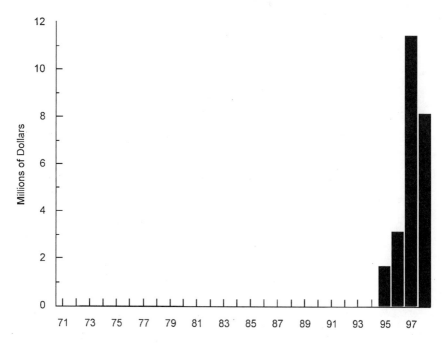

Figure A.3

References

Adair, H. S. 1927. Black pit of pecan. Am. Nut J. 26: 6–7.

Adair, H. S. 1932. Black pit of the pecan and some insects causing it. U.S.D.A. Circ. (N.S.) 234: 1–14.

Adams, D. B., and R. B. Chalfant. 1993. Vegetable insects, pp. 42–46. *In* R. M. McPherson and G. K. Douce (Eds.), Summary of losses from insect damage and costs of control in Georgia, 1992. Georgia Agric. Exp. Stn. Spec. Publ. 83: 1–55.

Adams, D. B., R. M. McPherson, D. C. Jones, and D. L. Horton. 1991. Soybean insects, pp. 30–31. *In* G. K. Douce and R. M. McPherson (Eds.), Summary of losses from insect damage and costs of control in Georgia, 1989. Ga. Agric. Exp. Stn. Spec. Publ. 70: 1–47.

Alcock, J. 1971. The behavior of a stinkbug, *Euschistus conspersus* Uhler (Hemiptera: Pentatomidae). Psyche 78: 215–228.

Aldrich, J. R. 1988a. Chemistry and biological activity of pentatomoid sex pheromones, pp. 417–431. *In* H. G. Cutler (Ed.), Biologically active natural products: potential use in agriculture. ACS Symposium Series 380, Washington, D. C. 483 pp.

Aldrich, J. R. 1988b. Chemical ecology of the Heteroptera. Ann. Rev. Entomol. 33: 211–238.

Aldrich, J. R. 1990. Dispersal of the southern green stink bug, *Nezara viridula* (L.) (Heteroptera: Pentatomidae), by Hurricane Hugo. Proc. Entomol. Soc. Washington 92: 757–759.

Aldrich, J. R. 1995a. Chemical communication in the true bugs and parasitoid exploitation, pp. 318–363. *In* R. T. Carde and W. J. Bell (Eds.), Chemical ecology of insects 2. Chapman and Hall, New York, NY. 433 pp.

Aldrich, J. R. 1995b. Testing the "new associations" biological control concept with a tachinid parasitoid (*Euclytia flava*). J. Chem. Ecol. 21: 1031–1042.

Aldrich, J. R. 1996. Sex pheromones in Homoptera and Heteroptera, pp. 199–233. *In* C. W. Schaefer (Ed.), Studies on hemipteran phylogeny. Proc. Thomas Say Publ. Entomol., Entomol. Soc. Am., Lanham, MD. 244 pp.

Aldrich, J. R. 1998. Status of semiochemical research on predatory Heteroptera, pp. 33–48. *In* M. Coll and J. R. Ruberson (Eds.), Predatory Heteroptera: their ecology and use in biological control. Proc. Thomas Say Publ. Entomol., Entomol. Soc. Am., Lanham, MD. 233 pp.

Aldrich, J. R., M. S. Blum, H. A. Lloyd, and H. M. Fales. 1978. Pentatomid natural products. Chemistry and morphology of the III-IV dorsal abdominal glands of adults. J. Chem. Ecol. 4: 161–172.

Aldrich, J. R., J. E. Oliver, W. R. Lusby, J. P. Kochansky, and J. A. Lockwood. 1987. Pheromone strains of the cosmopolitan pest, *Nezara viridula* (Heteroptera: Pentatomidae). J. Exp. Zool. 244: 171–175.

Aldrich, J. R., W. R. Lusby, B. E. Marron, K. C. Nicolaou, M. P. Hoffmann, and L. T. Wilson. 1989. Pheromone blends of green stink bugs and possible parasitoid selection. Naturwissenschaften 76: 173–175.

Aldrich, J. R., M. P. Hoffmann, J. P. Kochansky, W. R. Lusby, J. E. Eger, and J. A. Payne. 1991. Identification and attractiveness of a major pheromone component for Nearctic *Euschistus* spp. stink bugs (Heteroptera: Pentatomidae). Environ. Entomol. 20: 477–483.

Aldrich, J. R., H. Numata, M. Borges, F. Bin, G. K. Waite, and W. R. Lusby. 1993. Artifacts and pheromone blends from *Nezara* spp. and other stink bugs (Heteroptera: Pentatomidae). Z. Naturforschung 48: 73–79.

Aldrich, J. R., J. E. Oliver, W. R. Lusby, J. P. Kochansky, and M. Borges. 1994. Identification of male-specific volatiles from Nearctic and Neotropical stink bugs (Heteroptera: Pentatomidae). J. Chem. Ecol. 20: 1103–1111.

Aldrich, J. R., M. C. Rosi, and F. Bin. 1995. Behavioral correlates for minor volatile compounds from stink bugs (Heteroptera: Pentatomidae). J. Chem. Ecol. 21: 1907–1920.

Aldrich, J. R., J. W. Avery, C.-J. Lee, J. C. Graf, D. J. Harrison, and F. Bin. 1996. Semiochemistry of cabbage bugs (Heteroptera: Pentatomidae: *Eurydema* and *Murgantia*). J. Entomol. Sci. 31: 172–182.

Aldrich, S. R., W. O. Scott, and E. R. Leng. 1978. Modern corn production (2nd ed.). A & L Publ., Champaign, IL. 378 pp.

Ali, M., and M. A. Ewiess. 1977. Photoperiodic and temperature effects on rate of development and diapause in the green stink bug, *Nezara viridula* L. (Heteroptera: Pentatomidae). Zeits. Angew. Entomol. 84: 256–264.

Ali, M. A., A. M. Awadallah, and A. A. El-Rahman. 1979. A study on the phenology and ecology of the green stink bug *Nezara viridula* L. (Heteroptera: Pentatomidae). Zeits. Angew. Entomol. 88: 476–483.

All, J. N., and R. D. Hudson. 1993. Field corn insects, p. 9. *In* R. M. McPherson and G. K. Douce (Eds.), Summary of losses from insect damage and costs of control in Georgia, 1992. Georgia Agric. Exp. Stn. Spec. Publ. 83: 1–55.

Allen, C. T., P. Walden, B. Ree, and C. Turner. 1995. Use of southern pea trap crops to reduce pecan kernel damage in west Texas. Texas Agric. Ext. Serv. Result Demonstration Rep. 9 pp.

Altieri, M. A., J. W. Todd, E. W. Hauser, M. Patterson, G. A. Buchanan, and R. H. Walker. 1981. Some effects of weed management and row spacing on insect abundance in soybean fields. Prot. Ecol. 3: 339–343.

American Soybean Association. 1997. 1997 Soy stats: a reference guide to important soybean facts and figures. Am. Soybean Assoc., St. Louis, MO. 39 pp.

Anderson, R. M., and G. L. Teetes. 1996. Evaluation of insecticides for suppression of stink bugs on sorghum, 1995. Arthropod Manage. Tests 21: 285.

Anderson, R. M., and G. L. Teetes. 1997. Evaluation of insecticides for suppression of rice stink bug on sorghum, 1996. Arthropod Manage. Tests 22: 305–306.

Anderson, R. M., G. L. Teetes, and B. B. Pendleton. 1998. Evaluation of insecticides for suppression of rice stink bug on sorghum, 1997. Arthropod Manage. Tests 23: 274–275.

Annan, I. B., and M. K. Bergman. 1988. Effects of the onespotted stink bug (Hemiptera: Pentatomidae) on growth and yield of corn. J. Econ. Entomol. 81: 649–653.

Anonymous. 1945. Field crop insects in the Prairie Provinces. Line Elevators Farm Serv. (Winnipeg, Manitoba) Bull. 5: 1–64.

Anonymous. 1973. Small grains. Coop. Econ. Insect Rep. 23(18): 250.

Anonymous. 1974. Small grains. Coop. Econ. Insect Rep. 24(15): 223–224.

Anonymous. 1979. U.S.D.A. Coop. Plant Pest Rep. 4(11): 138, 147.

Anonymous. 1992. Wanted: a few good soldier (bugs). Science 257: 1049.

Anonymous. 1998. Insects and related pests, pp. 36–67. *In* L. L. Strand (Ed.), Integrated pest management for tomatoes (4th ed.). Univ. California, Div. Agric. & Nat. Res. Publ. 3274: 1–118.

Anonymous. 1999a. Sterling RESCUE!®. Pest control products. The RESCUE!® soldier bug attractor. http://www.rescue.com/products/soldierbug.html. 2 pp.

Anonymous. 1999b. Pheromone lures & traps. Stinkbug traps. http://www.sierraag.com/phlures/monitor/stinkbug.html. 2 pp.

Anonymous. 1999c. Stink bug traps. http://www.ipmtech.com/product/int_sb1.html. 1 p.

Apriyanto, D., J. D. Sedlacek, and L. H. Townsend. 1989a. Feeding activity of *Euschistus servus* and *E. variolarius* (Heteroptera: Pentatomidae) and damage to an early growth stage of corn. J. Kansas Entomol. Soc. 62: 392–399.

Apriyanto, D., L. H. Townsend, and J. D. Sedlacek. 1989b. Yield reduction from feeding by *Euschistus servus* and *E. variolarius* (Heteroptera: Pentatomidae) on stage V2 field corn. J. Econ. Entomol. 82: 445–448.

Arnaud, P. H., Jr. 1978. A host-parasite catalog of North American Tachinidae (Diptera). U.S.D.A. Misc. Publ. 1319. U.S. Government Printing Office, Washington, D.C. 860 pp.

Association Survey Resources. 1999. American crop protection association industry profile 1998. Assoc. Surv. Resources, LLC, Washington D.C. 22 pp. + appendix (9 pp.).

Awan, M. S., L. T. Wilson, and M. P. Hoffmann. 1990. Comparative biology of three geographic populations of *Trissolcus basalis* (Hymenoptera: Scelionidae). Environ. Entomol. 19: 387–392.

Bacheler, J. S., and D. W. Mott. 1995. Cotton insect management in North Carolina: IPM adoption following boll weevil eradication. Proc. 1995 Beltwide Cotton Conf. 2: 1067–1069.

Bailey, L. H., and E. Z. Bailey. 1976. Hortus third. A concise dictionary of plants cultivated in the United States and Canada. Macmillan Publ. Co., Inc., New York, NY. 1,290 pp.

Baldwin, J. L., D. J. Boethel, and R. Leonard. 1998. Control soybean insects 1998. Louisiana Coop. Ext. Serv. Publ. 2211: 1–16.

Barber, J. M. H. 1989. A proposal for integrated pest management of Say's stink bug on cultivated jojoba in southwestern Arizona. M.S. Thesis, Arizona State University, Tempe. 111 pp.

Barbour, K. S., J. R. Bradley, Jr., and J. S. Bacheler. 1988. Phytophagous stink bugs in North Carolina cotton: an evaluation of damage potential. Proc. Beltwide Cotton Production Res. Conf., pp. 280–282.

Barbour, K. S., J. R. Bradley, Jr., and J. S. Bacheler. 1990. Reduction in yield and quality of cotton damaged by green stink bug (Hemiptera: Pentatomidae). J. Econ. Entomol. 83: 842–845.

Barnett, W. W., B. E. Bearden, A. Berlowitz, C. S. Davis, J. L. Joos, and G. W. Morehead. 1976. True bugs cause severe pear damage. California Agric. 30(10): 20–23.

Basile, J. V., and R. B. Ferguson. 1964. Say stink bug destroys bitterbrush seed. J. Range Manage. 17: 153–154.

Baur, M. E., D. J. Boethel, M. L. Boyd, G. R. Bowers, M. O. Way, L. G. Heatherly, J. Rabb, and L. Ashlock. 2000. Arthropod populations in early soybean production systems in the mid-South. Environ. Entomol. 29: 312–328.

Beaulne, J. I. 1939. Parasites and predators reared at Quebec. Can. Entomol. 71: 120.

Bergman, M. K. 1999. Stink bugs (onespotted and brown), pp. 108–109. *In* K. L. Steffey, M. E. Rice, J. All, D. A. Andow, M. E. Gray, and J. W. Van Duyn (Eds.), Handbook of corn insects. Entomol. Soc. Am. Publ., Lanham, MD. 164 pp.

Beirne, B. P. 1972. Pest insects of annual crop plants in Canada. IV. Hemiptera — Homoptera, V. Orthoptera, VI. other groups. Mem. Entomol. Soc. Can. 85: 1–73.

Bernhardt, J. L., N. P. Tugwell, and R. N. Sharp. 1986. What is the rice quality imperfection called speck back? Texas Agric. Exp. Stn., Proc. 21st Rice Tech. Working Group, p. 91.

Bernhardt, J. L., N. P. Tugwell, R. N. Sharp, and W. H. Dodgen. 1987. Speck back — a new rice kernel imperfection. Arkansas Farm Res. 36(5): 10.

Bickenstaff, C. C., and J. L. Huggans. 1962. Soybean insects and related arthropods in Missouri. Missouri Agric. Exp. Stn. Res. Bull. 803: 1–51.

Blatchley, W. S. 1926. Heteroptera or true bugs of eastern North America with especial reference to the faunas of Indiana and Florida. Nature Publ. Co., Indianapolis, IN. 1116 pp.

Boethel, D. J. 1999. Assessment of soybean germplasm for multiple insect resistance, pp. 101–129. *In* S. L. Clement and S. S. Quisenberry (Eds.), Global plant genetic resources for insect-resistant crops. CRC Press LLC, Boca Raton, FL. 295 pp.

Boethel, D. J., J. S. Russin, A. T. Wier, M. B. Layton, J. S. Mink, and M. T. Boyd. 2000. Delayed maturity associated with southern green stink bug *Nezara viridula* (L.) (Heteroptera: Pentatomidae), injury at various soybean phenological stages. J. Econ. Entomol. 93: 707–712.

Borden, A. D., and H. F. Madsen. 1953. Control of stink bug on pears. Eradication of host plants in and near orchard and spring application of ground cover spray prove effective. California Agric. 7(2): 8.

Borden, A. D., H. F. Madsen, and A. H. Retan. 1952. A stink bug, *Euschistus conspersus*, destructive to deciduous fruits in California. J. Econ. Entomol. 45: 254–257.

Borges, M., and J. R. Aldrich. 1992. Instar-specific defensive secretions of stink bugs (Heteroptera: Pentatomidae). Experientia 48: 893–896.

Borges, M., P. C. Jepson, and P. E. Howse. 1987. Long-range mate location and close-range courtship behaviour of the green stink bug, *Nezara viridula* and its mediation by sex pheromones. Entomol. Exp. Appl. 44: 205–212.

Bowling, C. C. 1956. Control of the rice stink bug and grasshoppers on rice. Texas Agric. Exp. Stn. Prog. Rep. 1900: 1–3.

Bowling, C. C. 1962. Effect of insecticides on rice stink bug populations. J. Econ. Entomol. 55: 648–651.

Bowling, C. C. 1963. Cage tests to evaluate stink bug damage to rice. J. Econ. Entomol. 56: 197–200.

Bowling, C. C. 1967. Insect pests of rice in the United States, pp. 551–570. *In* M. D. Pathak (Ed.), The major insect pests of the rice plant. International Rice Research Institute. Johns Hopkins Press, Baltimore, MD. 729 pp.

Bowling, C. C. 1969. Estimation of rice stink bug populations on rice. J. Econ. Entomol. 62: 574–575.

Bowling, C. C. 1979. The stylet sheath as an indicator of feeding activity of the rice stink bug. J. Econ. Entomol. 72: 259–260.

Bowling, C. C. 1980. The stylet sheath as an indicator of feeding activity by the southern green stink bug on soybeans. J. Econ. Entomol. 73: 1–3.

Boyd, M. L., B. J. Fitzpatrick, and D. J. Boethel. 1996. Evaluation of selected insecticides for management of soybean arthropods in Louisiana, 1995. Arthropod Manage. Tests 21: 287–288.

Boyd, M. L., D. J. Boethel, B. R. Leonard, R. J. Habetz, L. P. Brown, and W. B. Hallmark. 1997. Seasonal abundance of arthropod populations on selected soybean varieties grown in early season production systems in Louisiana. Louisiana Agric. Exp. Stn. Bull. 860: 1–28.

Brako, L., A. Y. Rossman, and D. F. Farr. 1997. Scientific and common names of 7,000 vascular plants in the United States. Am. Phytopath. Soc. Publ., St. Paul, MN. 295 pp.

Brennan, B. M., F. Chang, and W. C. Mitchell. 1977. Physiological effects on sex pheromone communication in the southern green stink bug, *Nezara viridula*. Environ. Entomol. 6: 169–173.

Brett, C. H., and M. J. Sullivan. 1974. The use of resistance varieties and other cultural practices for control of insects on crucifers in North Carolina. North Carolina Agric. Exp. Stn. Bull. 449: 1–31.

Brewer, F. D., and W. A. Jones, Jr. 1985. Comparison of meridic and natural diets on the biology of *Nezara viridula* (Heteroptera: Pentatomidae) and eight other phytophagous Heteroptera. Ann. Entomol. Soc. Am. 78: 620–625.

Brezot, P., C. Malosse, K. Mori, and M. Renou. 1994. Bisabolene epoxides in sex pheromone in *Nezara viridula* (L.) (Heteroptera: Pentatomidae): role of *cis* isomer and relation to specificity of pheromone. J. Chem. Ecol. 20: 3133–3147.

Brison, F. R. 1986. Pecan culture. Texas Pecan Growers Assoc., Publ., College Station, TX. 297 pp.

Brook, T. S. 1953. Control of insects attacking rice in the field. Texas Agric. Exp. Stn. Prog. Rep. 1558: 1–3.

Brorsen, B. W., W. R. Grant, and M. E. Rister. 1984. Economic values of rice quality factors. Texas Agric. Exp. Stn. PR-4202: 1–26.

Brorsen, B. W., W. R. Grant, and M. E. Rister. 1988. Some effects of rice quality on rough rice prices. Southern J. Agric. Econ. 20: 131–140.

Brown, J. M. 1986. Forward, pp. xxi–xxii. *In* J. R. Mauney and J. McD. Stewart (Eds.), Cotton physiology. Cotton Found. Reference Book Ser., Memphis, TN. 786 pp.

Buchner, P. 1965. Endosymbiosis of animals with plant microorganisms. Interscience Publ., John Wiley & Sons, Inc., New York, NY. 909 pp.

Bugg, R. L., and J. D. Dutcher. 1989. Warm-season cover crops for pecan orchards: horticultural and entomological implications. Biol. Agric. Hortic. 6: 123–148.

Bundy, C. S., and R. M. McPherson. 2000a. Morphological examination of stink bug (Heteroptera: Pentatomidae) eggs on cotton and soybeans, with a key to genera. Ann. Entomol. Soc. Am. 93: 616–624.

Bundy, C. S., and R. M. McPherson. 2000b. Dynamics and seasonal abundance of stink bugs (Heteroptera: Pentatomidae) in a cotton-soybean ecosystem. J. Econ. Entomol. 93: 697–706.

Bundy, C. S., R. M. McPherson, and G. A. Herzog. 1998. Stink bugs in a cotton/soybean ecosystem: impact on quality and yield. Proc. 1998 Beltwide Cotton Conf. 2: 1172–1174.

Bundy, C. S., R. M. McPherson, and G.A Herzog. 2000. An examination of the external and internal signs of cotton boll damage by stink bugs (Heteroptera: Pentatomidae). J. Entomol. Sci. (in press).

Buschman, L. L., and W. H. Whitcomb. 1980. Parasites of *Nezara viridula* (Hemiptera: Pentatomidae) and other Hemiptera in Florida. Florida Entomol. 63: 154–162.

Buschman, L. L., H. N. Pitre, and H. F. Hodges. 1984. Soybean cultural practices: effects on populations of geocorids, nabids, and other soybean arthropods. Environ. Entomol. 13: 305–317.

Butler, G. D., Jr., and F. G. Werner. 1960. Pentatomids associated with Arizona crops. Arizona Agr. Exp. Stn. Tech. Bull. 140: 1–16.

Buxton, G. M., D. B. Thomas, and R. C. Froeschner. 1983. Revision of the species of the *sayi*-group of *Chlorochroa* Stal (Hemiptera: Pentatomidae). California Dept. Food and Agric., Occasional Papers Entomol. 29: 1–25.

Caffrey, D. J., and G. W. Barber. 1919. The grain bug. U.S.D.A. Bull. 779: 1–35.

Calhoun, D. S., J. E. Funderburk, and I. D. Teare. 1988. Soybean seed crude protein and oil levels in relation to weight, developmental time, and survival of southern green stink bug (Hemiptera: Pentatomidae). Environ. Entomol. 17: 727–729.

Callahan, P. S., R. Brown, and A. Dearman. 1960. Control of tomato insect pests. Louisiana Agric. 3(2): 4–5, 16.

Canerday, T. D. 1965. On the biology of the harlequin bug, *Murgantia histrionica* (Hemiptera: Pentatomidae). Ann. Entomol. Soc. Am. 58: 931–932.

Capone, T. A. 1995. Mutual preference for large mates in green stink bugs, *Acrosternum hilare* (Hemiptera: Pentatomidae). Anim. Behav. 49: 1335–1344.

Cassidy, T. P., and T. C. Barber. 1938. Hemipterous cotton insects of Arizona and their economic importance and control. U.S.D.A. Bur. Entomol. E-439: 1–16.

Cassidy, T. P., and T. C. Barber. 1939. Hemipterous insects of cotton in Arizona: their economic importance and control. J. Econ. Entomol. 32: 99–104.

Cassidy, T. P., and T. C. Barber. 1940. Investigations in control of hemipterous cotton insects in Arizona by the use of insecticides. U.S.D.A. Bur. Entomol. E-506: 1–20.

Castro, G. D. 1985. Insect problems on sorghum in Mexico. International Crops Research Institute for the Semi-arid Tropics, International Sorghum Entomology Workshop (Texas A&M Univ.) Proceed.: 83–87.

Chalfant, R. B. 1973. Chemical control of the southern green stinkbug, tomato fruitworm and potato aphid on vining tomatoes in southern Georgia. J. Georgia Entomol. Soc. 8: 279–283.

Chalfant, R. B. 1985. Entomological research on cowpea pests in the USA, pp. 265–271. *In* S. R. Singh and K. O. Rachie (Eds.), Cowpea research, production and utilization. John Wiley & Sons, New York, NY. 460 pp.

Chalfant, R. B., and J. R. Young. 1982. Chemigation, or application of insecticide through overhead sprinkler irrigation systems, to manage insect pests infesting vegetable and agronomic crops. J. Econ. Entomol. 75: 237–241.

Chambliss, C. E. 1920. Prairie rice culture in the United States. U.S.D.A. Farmers' Bull. 1092: 1–26.

Chandler, S. C. 1943. Can we control cat-facing and curculio in Illinois? Trans. Illinois State Hortic. Soc. 77: 493–505.

Chandler, S. C. 1950. Peach insects of Illinois and their control. Illinois Nat. Hist. Surv. Circ. 43: 1–63.

Chandler, S. C. 1955. Biological studies of peach catfacing insects in Illinois. J. Econ. Entomol. 48: 473–475.

Chandler, S. C., and W. P. Flint. 1939. Controlling peach insects in Illinois. Illinois Nat. Hist. Surv. Circ. 33: 1–40.

Chang, T. and E. A. Bardenas. 1965. The morphology and varietal characteristics of the rice plant. Int. Rice Res. Inst., Laguna, Philippines, Tech. Bull. 4: 1–40.

Cherry, E. T. 1973. Economic threshold studies of stink bugs on soybeans. Tennessee Farm Home Sci. Prog. Rep. 87: 8–9.

Cherry, R., D. Jones, and C. Deren. 1998. Establishment of a new stink bug pest, *Oebalus ypsilongriseus* (Hemiptera: Pentatomidae) in Florida rice. Florida Entomol. 81: 216–220.

Chittenden, F. H. 1908. The harlequin cabbage bug (*Murgantia histrionica* Hahn.). U.S.D.A. Bur. Entomol. Circ. 103: 1–10.

Chittenden, F. H. 1920. Harlequin cabbage bug and its control. U.S.D.A. Farmers' Bull. 1061: 1–13.

Chyen, D., M. E. Wetzstein, R. M. McPherson, and W. D. Givan. 1992. An economic evaluation of soybean stink bug control alternatives for the southeastern United States. Southern J. Agric. Econ. 24: 83–94.

Clair, D. J., and J. E. McPherson. 1980. Effects of temperature on development and reproduction in *Euschistus tristigmus tristigmus* (Hemiptera: Pentatomidae) with notes on reproductive behavior. Trans. Illinois State Acad. Sci. 73(2): 80–92.

Clancy, D. W. 1946. Natural enemies of some Arizona cotton insects. J. Econ. Entomol. 39: 326–328.

Clarke, R. G, and G. E. Wilde. 1970a. Association of the green stink bug and the yeast-spot disease organism of soybeans. 1. Length of retention, effect of molting, isolation from feces and saliva. J. Econ. Entomol. 63: 200–204.

Clarke, R. G., and G. E. Wilde. 1970b. Association of the green stink bug and the yeast-spot disease organism of soybeans. II. Frequency of transmission to soybeans, transmission from insect to insect, isolation from field population. J. Econ. Entomol. 63: 355–357.

Clarke, R. G., and G. E. Wilde. 1971. Association of the green stink bug and the yeast-spot disease organism of soybeans. III. Effect on soybean quality. J. Econ. Entomol. 64: 222–223.

Clower, D. F. 1958. Damage to corn by the southern green stink bug. J. Econ. Entomol. 51: 471–473.

Clower, D. F., and T. E. Hankins. 1960. Stinkbugs — a new menace to Louisiana soybeans, pp. 7–8. In W. T. Spink (Ed.), Insect conditions in Louisiana 1960. Louisiana State University 3: 1–58.

Cogburn, R. R., and M. O. Way. 1991. Relationship of insect damage and other factors to the incidence of speckback, a site-specific lesion on kernels of milled rice. J. Econ. Entomol. 84: 987–995.

Cokl, A., M. Gogala, and M. Jez. 1972. The analysis of the acoustic signals of the bug *Nezara viridula*. Biol. Vestnik 20: 47–53.

Cokl, A., M. Gogala, and A. Blazevic. 1978. Principles of sound recognition in three pentatomide bug species (Heteroptera). Biol. Vestnik 26: 81–94.

Correa-Ferreira, B. S., and F. Moscardi. 1996. Biological control of soybean stink bugs by inoculative releases of *Trissolcus basalis*. Entomol. Exp. Appl. 79: 1–7.

Crocker, T. F. 1984a. The history of pecans, p. 5. In H C Ellis, P. F. Bertrand, and T. F. Crocker (Eds.), Pecan pest management in the Southeast. Georgia Coop. Ext. Serv. Misc. Publ. 176: 1–62.

Crocker, T. F. 1984b. Pecan growth, development and nut drop, pp. 5–7. In H C Ellis, P. F. Bertrand, and T. F. Crocker (Eds.), Pecan pest management in the Southeast. Georgia Coop. Ext. Serv. Misc. Publ. 176: 1–62.

Cronholm, G., A. Knutson, R. Parker, G. Teetes, and B. Pendelton. 1998. Managing insect and mite pests of Texas sorghum. Texas Agric. Ext. Serv. Publ. B-1220: 1–26.

Cullen, E. M., and F. G. Zalom. 2000. Phenology-based field monitoring for consperse stink bug (Hemiptera: Pentatomidae) in processing tomatoes. Environ. Entomol. 29: 560–567.

Culliney, T. W. 1985. Predation on the imported cabbageworm *Pieris rapae* by the stink bug *Euschistus servus euschistoides* (Hemiptera: Pentatomidae). Can. Entomol. 117: 641–642.

Dahms, R. G. 1942. Rice stinkbug as a pest of sorghums. J. Econ. Entomol. 35: 945–946.

Daniels, L. B. 1939. Appearance of a new potato disease in northeastern Colorado. Science 90(2334): 273.

Daniels, L. B. 1941. Colorado potato pests. Colorado Exp. Stn. Bull. 465: 1–28.

da Silva, P., and K. Daane. 1995. Stinkbugs and leaffooted bugs, pp. 126–128. In L. Ferguson (Ed.), Pistachio production 1995. Univ. California, Davis, Center for Fruit and Nut Crop Research and Information. 160 pp.

Daugherty, D. M. 1967. Pentatomidae as vectors of yeast-spot disease of soybeans. J. Econ. Entomol. 60: 147–152.

Daugherty, D. M., and J. E. Foster. 1966. Organism of yeast-spot disease isolated from rice damaged by rice stink bug. J. Econ. Entomol. 59: 1282–1283.

Daugherty, D. M., M. H. Neustadt, C. W. Gehrke, L. E. Cavanah, L. F. Williams, and D. E. Green. 1964. An evaluation of damage to soybeans by brown and green stink bugs. J. Econ. Entomol. 57: 719–722.

Davis, C. J. 1964. The introduction, propagation, liberation, and establishment of parasites to control *Nezara viridula* variety *smaragdula* (Fabricius) in Hawaii (Heteroptera: Pentatomidae). Proc. Hawaiian Entomol. Soc. 18: 369–375.

Davis, C. J. 1966. Progress in the biological control of the southern green stink bug, *Nezara viridula* variety *smaragdula* (Fabricius) in Hawaii (Heteroptera: Pentatomidae). Mushi 39: 9–16.

Davis, W. T. 1925. Note on *Podops cinctipes* and *Solubea pugnax*. Bull. Brooklyn Entomol. 20: 147.

DeCoursey, R. M., and C. O. Esselbaugh. 1962. Descriptions of the nymphal stages of some North American Pentatomidae (Hemiptera-Heteroptera). Ann. Entomol. Soc. Am. 55: 323–342.

Deighan, J., R. M. McPherson, and F. W. Ravlin. 1985. Comparison of sweep-net and ground-cloth sampling methods for estimating arthropod densities in different soybean cropping systems. J. Econ. Entomol. 78: 208–212.

Deitz, L. L., J. W. Van Duyn, J. R. Bradley, Jr., R. L. Rabb, W. M. Brooks, and R. E. Stinner. 1976. A guide to the identification and biology of soybean arthropods in North Carolina. North Carolina Agric. Exp. Stn. Tech. Bull. 238: 1–264.

Delaplane, K. S. (Ed.). 1995. Georgia pest control handbook. Georgia Coop. Ext. Serv. Spec. Bull. 28: 1–527.

Del Vecchio, M. C. and J. Grazia. 1992a. Obtencao de posturas de *Oebalus ypsilongriseus* (De Geer, 1773) em Laboratorio (Heteroptera: Pentatomidae). Anais Soc. Entomol. Brasil 21: 367–373.

Del Vecchio, M. C. and J. Grazia. 1992b. Estudo dos imaturos *Oebalus ypsilongriseus* (De Geer, 1773): I - Descricao do ovo e desenvolvimento embrionario (Heteroptera: Pentatomidae). Anais Soc. Entomol. Brasil 21: 375–382.

Del Vecchio, M. C. and J. Grazia. 1993a. Estudo dos imaturos de *Oebalus ypsilongriseus* (De Geer, 1773): III - Duracao e mortalidade dos estagios de ovo e ninfa (Heteroptera: Pentatomidae). Anais Soc. Entomol. Brasil 22: 121–129.

Del Vecchio, M. C., and J. Grazia. 1993b. Estudo dos imaturos de *Oebalus ypsilongriseus* (De Geer, 1773): II - Descricao das ninfas (Heteroptera: Pentatomidae). Anais Soc. Entomol. Brasil 22: 109–120.

Del Vecchio, M. C., J. Grazia, and G. S. Albuquerque. 1994. Dimorfismo sazonal em *Oebalus ypsilongriseus* (De Geer, 1773) (Hemiptera, Pentatomidae) e uma nova sinonimia. Revta Bras. Entomol. 38: 101–108.

Demaree, J. B. 1922. Kernel-spot of the pecan and its cause. U.S.D.A. Bull. 1102: 1–15.

DeWitt, N. B., and G. L. Godfrey. 1972. The literature of arthropods associated with soybeans. II. A bibliography of the southern green stink bug *Nezara viridula* (Linneaus) [sic] (Hemiptera: Pentatomidae). Illinois Nat. Hist. Surv. Biol. Notes 78: 1–23.

Doggett, H. 1988. Sorghum (2nd ed.). John Wiley & Sons, Inc., New York, NY. 512 pp.

Douglas, W. A. 1939. Studies of rice stinkbug populations with special reference to local migration. J. Econ. Entomol. 32: 300–303.

Douglas, W. A., and J. W. Ingram. 1942. Rice-field insects. U.S.D.A. Circ. 632: 1–32.

Douglas, W. A., and E. C. Tullis. 1950. Insects and fungi as causes of pecky rice. U.S.D.A. Tech. Bull. 1015: 1–20.

Drake, C. J. 1920. The southern green stink-bug in Florida. Florida State Plant Board Quarterly Bull. 4: 41–94.

Drees, B. M. 1983. Rice insect management. Texas Agric. Ext. Serv. Publ. B-1445: 1–12.

Drees, B. M., and F. W. Plapp. 1986. Toxicity of carbaryl and methyl parathion to populations of rice stink bugs, *Oebalus pugnax* (Fabricius). Texas Agric. Exp. Stn. PR-4415: 1–7.

Drees, B. M., and M. E. Rice. 1990. Population dynamics and seasonal occurrence of soybean insect pests in southeastern Texas. Southwestern Entomol. 15: 49–56.

Drickamer, L. C., and J. E. McPherson. 1992. Comparative aspects of mating behavior patterns in six species of stink bugs (Heteroptera: Pentatomidae). Great Lakes Entomol. 25: 287–295.

Duncan, R. G., and J. R. Walker. 1968. Some effects of the southern green stink bug on soybeans. Louisiana Agric. 12(2): 10–11.

Dutcher, J. D. 1984. Stink bugs and leaf-footed bugs, p. 28. In H C Ellis, P. F. Bertrand, and T. F. Crocker (Eds.), Pecan pest management in the Southeast. Georgia Coop. Ext. Serv. Misc. Publ. 176: 1–62.

Dutcher, J. D. 1987. Two difficult pecan insect problems - aphids and kernel-feeding hemipterans. Proc. Southeastern Pecan Growers Assoc. 80: 73–75.

Dutcher, J. D. 1989. Stink bugs and leaffooted bugs, p. 122. In W. D. Goff, J. R. McVay, and W. S. Gazaway (Eds.), Pecan production in the southeast: a guide for growers. Alabama Coop. Ext. Circ. ANR-459: 1–222. (reprinted 1996).

Dutcher, J. D. 1990. Kernel damage may require more than insecticides. The Pecan Grower 1(1): 13–15.

Dutcher, J. D., and J. W. Todd. 1983. Hemipteran kernel damage of pecan, pp. 1–11. In J. A. Payne (Ed.), Pecan pest management - are we there? Misc. Publ. Entomol. Soc. Am. 13: 1–140.

Dyby, S. D., and R. I. Sailer. 1999. Impact of low-level radiation on fertility and fecundity of Nezara viridula (Hemiptera: Pentatomidae). J. Econ. Entomol. 92: 945–953.

Eger, J. E., Jr., and J. R. Ables. 1981. Parasitism of Pentatomidae by Tachinidae in South Carolina and Texas. Southwest. Entomol. 6: 28–33.

Ehler, L. E. 2000. Farmscape ecology of stink bugs in northern California. Mem. Thomas Say Publ. Entomol., Entomol. Soc. Am. Press, Lanham, MD. 59 pp.

Eigenbrode, S. D., J. T. Trumble, W. G. Carson, and K. K. White. 1993. Susceptibility of fresh market tomato varieties to insect pests in southern California, 1992. Insecticide and Acaricide Tests 18: 172–173.

Elliott, N. C., G. L. Hein, and B. M. Shepard. 1994. Sampling arthropod pests of wheat and rice, pp. 627–666. In L. P. Pedigo and G. D. Buntin (Eds.), Handbook of sampling methods for arthropods in agriculture. CRC Press, Inc., Boca Raton, FL. 714 pp.

Ellis, H C. 1984. Techniques of pest management, pp. 8–12. In H C Ellis, P. F. Bertrand, and T. F. Crocker (Eds.), Pecan pest management in the Southeast. Georgia Coop. Ext. Serv. Misc. Publ. 176: 1–62.

Elsey, K. D. 1993. Cold tolerance of the southern green stink bug (Heteroptera: Pentatomidae). Environ. Entomol. 22: 567–570.

English-Loeb, G. M., and B. D. Collier. 1987. Nonmigratory movement of adult harlequin bugs Murgantia histrionica (Hemiptera: Pentatomidae) as affected by sex, age and host plant quality. Am. Midland Nat. 118: 189–197.

Esselbaugh, C. O. 1946. A study of the eggs of the Pentatomidae (Hemiptera). Ann. Entomol. Soc. Am. 39: 667–691.

Esselbaugh, C. O. 1948. Notes on the bionomics of some midwestern Pentatomidae. Entomol. Am. 28: 1–73.

Essig, E. O. 1926. Insects of western North America. Macmillan Co., New York, NY. 1035 pp.

Faust, W., and R. H. Strang (Eds.). 1983. Important weeds of the world (scientific and common names, synonyms, and WSSA approved computer codes). Bayer AG, Agrochem. Div. Publ., Leverkusen, Germany. 711 pp.

Fehr, W. R. and C. E. Caviness. 1977. Stages of soybean development. Iowa Coop. Ext. Serv. Spec. Rep. 80: 1–12.

Fehr, W. R., C. E. Caviness, D. T. Burmood, and J. S. Pennington. 1971. Stage of development descriptions for soybeans, *Glycine max* (L.) Merrill. Crop Sci. 11:929–931.

Ferguson, R. B., M. M. Furniss, and J. V. Basile. 1963. Insects destructive to bitterbrush flowers and seeds in southwestern Idaho. J. Econ. Entomol. 56: 459–462.

Fernald, M. L. 1950. Gray's manual of botany (8th ed.). American Book Co., New York, NY. 1632 pp.

Fery, R. L. 1990. The cowpea: production, utilization, and research in the United States, pp. 197–222. *In* J. Janic (Ed.), Horticultural reviews. Volume 12. Timber Press, Inc., Portland OR. 509 pp.

Fery, R. L., and J. M. Schalk. 1981. An evaluation of damage to southernpeas (*Vigna unguiculata* (L.) Walp.) by leaffooted and southern green stink bugs. HortScience 16: 286.

Fery, R. L., and J. M. Schalk. 1984. Southern green stink bug: identification of resistance in cowpea. HortScience 19: 211.

Fish, J., and J. Alcock. 1973. The behavior of *Chlorochroa ligata* (Say) and *Cosmopepla bimaculata* (Thomas), (Hemiptera: Pentatomidae). Entomol. News 84: 260–268.

Fitzpatrick, B. J., and D. J. Boethel. 1998. Evaluation of insecticides for green stink bug and brown stink bug control on soybean, 1997. Arthropod Manage. Tests 23: 280–281.

Fontes, E. M. G., D. H. Habeck, and F. Slansky, Jr. 1994. Phytophagous insects associated with goldenrods (*Solidago* spp.) in Gainesville, Florida. Florida Entomol. 77: 209–221.

Foot, K. and E. C. Strobell. 1914. Results of crossing *Euschistus variolarius* and *Euschistus servus* with reference to the inheritance of an exclusively male character. J. Linn. Soc. London (Zool.) 32: 337–373.

Forbes, R. S., and L. Daviault. 1964. The biology of the mountain-ash sawfly, *Pristiphora geniculata* (Htg.) (Hymenoptera: Tenthredinidae) in eastern Canada. Can. Entomol. 96: 1117–1133.

Forbes, S. A. 1905. Noxious and beneficial insects of the State of Illinois. A monograph of insect injuries to Indian corn. Part II. Twenty-third Report of the State Entomologist. Twelfth Report of S. A. Forbes. 273 pp.

Foster, J. E., and D. M. Daugherty. 1969. Isolation of the organism causing yeast-spot disease from the salivary system of the green stink bug. J. Econ. Entomol. 62: 424–427.

Foster, R. E., R. H. Cherry, and D. B. Jones. 1986. Distribution of rice stink bugs in Florida rice fields. Florida Agric. Coop. Ext. Serv., Belle Glade Everglades Res. Educ. Center, Res. Rep. EV- 1986–6: 11–12.

Foster, R. E., R. H. Cherry, and D. B. Jones. 1989. Spatial distribution of the rice stink bug (Heteroptera: Pentatomidae) in Florida rice. J. Econ. Entomol. 82: 507–509.

Franqui, R. A., A. Pantoja, and S. Medina-Gaud. 1988. Host plants of pentatomids affecting rice fields in Puerto Rico. J. Agric. (Univ. Puerto Rico) 72: 365–369.

Froeschner, R. C. 1941. Contributions to a synopsis of the Hemiptera of Missouri, Pt. 1. Scutelleridae, Podopidae, Pentatomidae, Cydnidae, Thyreocoridae. Am. Midland Nat. 26: 122–146.

Froeschner, R. C. 1988. Family Pentatomidae Leach, 1815. The stink bugs, pp. 544–597. *In* T. J. Henry and R. C. Froeschner (Eds.), Catalog of the Heteroptera, or true bugs, of Canada and the continental United States. E. J. Brill, New York, NY. 958 pp.

Fryar, E. O., L. D. Parsch, S. H. Holder, and N. P. Tugwell. 1986a. The economics of controlling peck in Arkansas rice. Arkansas Farm Res. 35(3): 7.

Fryar, E. O., L. D. Parsch, S. H. Holder, and N. P. Tugwell. 1986b. Reducing peck: is it worth it? Texas Agric. Exp. Stn., Proc. 21st Rice Tech. Working Group, pp. 115–116.

Fuchs, T. W., H. A. Turney, J. G. Thomas, and G. L. Teetes. 1988. Managing insect and mite pests of Texas sorghum. Texas Agric. Ext. Serv. B-1220: 1–16.

Funderburk, J. E., D. L. Wright, and I. D. Teare. 1990. Preplant tillage effects on population dynamics of soybean insect pests. Crop Sci. 30: 686–690.

Funderburk, J. E., I. D. Teare, and F. M. Rhoads. 1991. Population dynamics of soybean insect pests vs. soil nutrient levels. Crop Sci. 31: 1629–1633.

Funderburk, J., R. McPherson, and D. Buntin. 1999. Soybean insect management, pp. 273–290. *In* L. G. Heatherly and H. F. Hodges (Eds.), Soybean production in the Midsouth. CRC Press LLC, Boca Raton, FL. 394 pp.

Furth, D. G. 1974. The stink bugs of Ohio (Hemiptera: Pentatomidae). Bull. Ohio Biol. Surv. (N.S.) 5(1): 1-60.

Ganyard, M. C., J. R. Brazzel, and J. F. Kearney. 1979. Progress report on the boll weevil eradication trial. Proc. 1979 Beltwide Cotton Production-Mechanization Conf., pp. 27–29.

Garman, H. 1891. *Oebalus pugnax* an enemy of grasses. Psyche 6: 61.

Genung, W. G., V. E. Green, Jr., and C. Wehlburg. 1964. Inter-relationship of stinkbugs and diseases to Everglades soybean production. Soil Crop Sci. Soc. Florida Proc. 24: 131–137.

Genung, W. G., G. H. Snyder, and V. E. Green, Jr.. 1979. Rice field insects in the Everglades. Their biology and ecology. Florida Agric Coop. Ext. Serv., Belle Glade Everglades Res. Educ. Center, Res. Rep. EV-1979-7: 1–28 + 14 figures.

Gifford, J. R., B. F. Oliver, and G. B. Trahan. 1968a. New insect pests of rice in Louisiana. Louisiana Agric. Exp. Stn., Rice Exp. Stn., Annu. Prog. Rep. 60: 174–181.

Gifford, J. R., B. F. Oliver, and G. B. Trahan. 1968b. Control of the rice stink bug, *Oebalus pugnax* (F.) and long-horned grasshoppers in rice. Louisiana Agric. Exp. Stn., Rice Exp. Stn., Annu. Prog. Rep. 60: 158–173.

Gilby, A. R., and D. F. Waterhouse. 1965. The composition of the scent of the green vegetable bug, *Nezara viridula*. Proc. Roy. Soc. London, Ser. B., Biol. Sci. 162: 105–120.

Gilby, A. R., and D. F. Waterhouse. 1967. Secretions from the lateral scent glands of the green vegetable bug, *Nezara viridula*. Nature 216: 90–91.

Gill, J. B. 1924. Important pecan insects and their control. U.S.D.A. Farmers' Bull. 1364: 1–48.

Gillette, C. P. 1904. Report of the entomologist. I. Some of the more important insects of 1903 and an annotated list of Colorado Orthoptera, pp. 1–56. Colorado Agric. Exp. Stn. Bull. 94 (Tech. Ser. 6): 1–85.

Gilman, D. F., R. M. McPherson, L. D. Newsom, D. C. Herzog, and C. Williams. 1982. Resistance in soybeans to the southern green stink bug. Crop Sci. 22: 573–576.

Girault, A. A. 1907. Hosts of insect egg-parasites in North and South America. Psyche 14: 27–39.

Goeden, R. D., and D. W. Ricker. 1968. The phytophagous insect fauna of Russian thistle (*Salsola kali* var. *tenuifolia*) in southern California. Ann. Entomol. Soc. Am. 61: 67–72.

Grant, W. R., M. E. Rister, and B. W. Brorsen. 1986a. Rice quality factors: implications for management decisions. Texas Agric. Exp. Stn. Publ. B-1541: 1–77.

Grant, W. R., M. E. Rister, and B. W. Brorsen. 1986b. Some effects of rice quality on rough rice prices. Texas Agric. Exp. Stn., Proc. 21st Rice Tech. Working Group, p. 115.

Grazia, J., M. C. Del Vecchio, F. M. P. Balestieri, and Z. A. Ramiro. 1980. Estudo das ninfas de pentatomideos (Heteroptera) que vivem sobre soja (*Glycine max* [L.] Merrill): I - *Euchistus* [sic] *heros* (Fabricius, 1798) e *Piezodorus guildinii* (Westwood, 1837). Anais Soc. Entomol. Brasil 9: 39–51.

Green, V. E., Jr., W. H. Thames, Jr., A. E. Kretschmer, Jr., A. L. Craig, and E. C. Tullis. 1954. Rice investigations. Florida Agric. Exp. Stn. Ann. Rep., pp. 235–236.

Greene, J. K., S. G. Turnipseed, and M. J. Sullivan. 1998. Managing stink bugs in *Bt* cotton. Proc. 1998 Beltwide Cotton Conf. 2: 1174–1177.

Greene, J. K., S. G. Turnipseed, M. J. Sullivan, and G. A. Herzog. 1999. Boll damage by southern green stink bug (Hemiptera: Pentatomidae) and tarnished plant bug (Hemiptera: Miridae) caged on transgenic *Bacillus thuringiensis* cotton. J. Econ. Entomol. 92: 941–944.

Griffin, R. P., and M. F. R. Khan. 1998a. Efficacy of insecticides on tomato insects, 1996. Arthropod Manage. Tests 23: 153–154.

Griffin, R. P., and M. F. R. Khan. 1998b. Efficacy of insecticides on tomato insects, 1997. Arthropod Manage. Tests 23: 154–155.

Grigarick, A. A. 1984. General problems with rice invertebrate pests and their control in the United States. Prot. Ecol. 7: 105–114.

Guillebeau, P. 1998. Names, classification and toxicity of pesticides, p. 529. *In* P. Guillebeau (Ed.), Georgia pest control handbook. Georgia Ext. Serv. Spec. Bull. 28: 1–573.

Hall, D. G., IV, and G. L. Teetes. 1980. Damage to sorghum seed by four common bugs. Texas Agric. Exp. Stn. PR-3647: 1–8.

Hall, D. G., IV, and G. L. Teetes. 1981. Alternate host plants of sorghum panicle-feeding bugs in southeast central Texas. Southwestern Entomol. 6: 220–228.

Hall, D. G., IV, and G. L. Teetes. 1982a. Damage by rice stink bug to grain sorghum. J. Econ. Entomol. 75: 440–445.

Hall, D. G., IV, and G. L. Teetes. 1982b. Damage to grain sorghum by southern green stink bug, conchuela, and leaffooted bug. J. Econ. Entomol. 75: 620–625.

Hall, D. G., IV, and G. L. Teetes. 1982c. Yield loss-density relationships of four species of panicle-feeding bugs in sorghum. Environ. Entomol. 11: 738–741.

Hall, D. G., IV, G. L. Teetes, and C. E. Hoelscher. 1983. Suggested guidelines for controlling panicle-feeding bugs in Texas sorghum. Texas Agric. Ext. Serv. B-1421. 7 pp.

Hallman, G. J., C. G. Morales, and M. C. Duque. 1992. Biology of *Acrosternum marginatum* (Heteroptera: Pentatomidae) on common beans. Florida Entomol. 75: 190–196.

Hamer, J., and J. Jarratt. 1983. Rice insect control. Mississippi Coop. Ext. Serv. 888. 4 pp.

Hammond, R. B., R. A. Higgins, T. P. Mack, L. P. Pedigo, and E. J. Bechinski. 1991. Soybean pest management, pp. 341–472. *In* D. Pimentel (Ed.), CRC handbook of pest management in agriculture (2nd ed.). Volume III. CRC Press, Inc., Boca Raton, FL. 749 pp.

Harper, J. D., R. M. McPherson, and M. Shepard. 1983. Geographical and seasonal occurrence of parasites, predators and entomopathogens, pp. 7–19. *In* H. N. Pitre (Ed.), Natural enemies of arthropod pests in soybean. South Carolina Agric. Exp. Stn. (Southern Coop. Ser.) Bull. 285: 1–90.

Harper, J. K., M. E. Rister, J. W. Mjelde, B. M. Drees, and M. O. Way. 1990. Factors influencing the adoption of insect management technology. Am. J. Agric. Econ. 72: 997–1005.

Harper, J. K., M. O. Way, B. M. Drees, M. E. Rister, and J. W. Mjelde. 1993. Damage function analysis for the rice stink bug (Hemiptera: Pentatomidae). J. Econ. Entomol. 86: 1250–1258.

Harper, J. K., J. W. Mjelde, M. E. Rister, M. O. Way, and B. M. Drees. 1994. Developing flexible economic thresholds for pest management using dynamic programming. J. Agric. Appl. Econ. 26: 134–147.

Harris, R. H., L. D. Sibbitt, J. A. Munro, and H. S. Telford. 1941. The effect of the green grain bug upon the milling and baking quality of wheat. North Dakota Agric. Exp. Stn. Bimonthly Bull. 3(5): 10–14.

Harris, V. E., and J. W. Todd. 1980a. Male-mediated aggregation of male, female and 5th-instar southern green stink bugs and concomitant attraction of a tachinid parasite, *Trichopoda pennipes*. Entomol. Exp. Appl. 27: 117–126.

Harris, V. E., and J. W. Todd. 1980b. Temporal and numerical patterns of reproductive behavior in the southern green stink bug, *Nezara viridula* (Hemiptera: Pentatomidae). Entomol. Exp. Appl. 27: 105–116.

Harris, V. E., and J. W. Todd. 1980c. Duration of immature stages of the southern green stink bug, *Nezara viridula* (L.), with a comparative review of previous studies. J. Georgia Entomol. Soc. 15: 114–124.

Harris, V. E., and J. W. Todd. 1980d. Comparative fecundity, egg fertility and hatch among wild-type and three laboratory-reared generations of the southern green stink bug, *Nezara viridula* (L.) (Hemiptera: Pentatomidae). J. Georgia Entomol. Soc. 15: 245–252.

Harris, V. E., and J. W. Todd. 1981a. Validity of estimating percentage parasitization of *Nezara viridula* populations by *Trichopoda pennipes* using parasite-egg presence on host cuticle as the indicator. J. Georgia Entomol. Soc. 16: 505–510.

Harris, V. E., and J. W. Todd. 1981b. Rearing the southern green stink bug, *Nezara viridula*, with relevant aspects of its biology. J. Georgia Entomol. Soc. 16: 203–210.

Harris, V. E., J. W. Todd, J. C. Webb, and J. C. Benner. 1982. Acoustical and behavioral analysis of the songs of the southern green stink bug, *Nezara viridula*. Ann. Entomol. Soc. Am. 75: 234–249.

Harris, V. E., J. W. Todd, and B. G. Mullinix. 1984. Color change as an indicator of adult diapause in the southern green stink bug, *Nezara viridula*. J. Agric. Entomol. 1: 82–91.

Hayes, W. P. 1922. A preliminary list of insects of the sorghum field. Trans. Kansas Acad. Sci. 30: 235–240.

Heatherly, L. G., and G. Bowers (Eds.). 1998. Early soybean production system handbook. United Soybean Board. Mississippi State Univ., Office Agric. Communications. 26 pp.

Helm, R. W. 1954. Pecky rice damage caused by rice stink bug during 1953. Rice Journal 57(8): 29.

Higley, L. G. 1994. Soybean pest management procedures, p. 111. *In* L. G. Higley and D. J. Boethel (Eds.), Handbook of soybean insect pests. Entomol. Soc. Am. Publ., Lanham, MD. 136 pp.

Hills, O. A. 1941. Isolation-cage studies of certain hemipterous and homopterous insects on sugar beets grown for seed. J. Econ. Entomol. 34: 756–760.

Hills, O. A. 1943. Comparative ability of several species of lygus and the Say stinkbug to damage sugar beets grown for seed. J. Agric. Res. 67: 389–394.

Hills, O. A., and K. B. McKinney. 1946. Damage by *Euschistus impictiventris* and *Chlorochroa sayi* to sugar beets grown for seed. J. Econ. Entomol. 39: 335–337.

Hodson, A. C., and E. F. Cook. 1960. Long-range aerial transport of the harlequin bug and the greenbug into Minnesota. J. Econ. Entomol. 53: 604–608.

Hoffmann, M. P., L. T. Wilson, and F. G. Zalom. 1987a. The southern green stink bug, *Nezara viridula* Linnaeus (Heteroptera: Pentatomidae); new location. Pan-Pac. Entomol. 63: 333.

Hoffmann, M. P., L. T. Wilson, and F. G. Zalom. 1987b. Control of stink bugs in tomatoes. California Agric. 41(5–6): 4–6.

Hoffmann, M. P., N. A. Davidson, L. T. Wilson, L. E. Ehler, W. A. Jones, and F. G. Zalom. 1991. Imported wasp helps control southern green stink bug. California Agric. 45(3): 20–22.

Hoffmann, W. E. 1935. The foodplants of *Nezara viridula* Linn. (Hem. Pent.). Proc. Internatl. Congr. Entomol. 6: 811–816.

Hogmire, H. W., and T. Winfield. 1998. Insecticide evaluation, 1997. Arthropod Manage. Tests 23: 43–44.

Hokkanen, H. M. T. 1991. Trap cropping in pest management. Annu. Rev. Entomol. 36: 119–138.

Hokyo, N., and K. Kiritani. 1962. Sampling design for estimating the population of the southern green stink bug, *Nezara viridula* (Pentatomidae, Hemiptera) in the paddy field. Japanese J. Ecol. 12: 228–235.

Hollay, M. E., C. M. Smith, and J. F. Robinson. 1987. Structure and formation of feeding sheaths of rice stink bug (Heteroptera: Pentatomidae) on rice grains and their association with fungi. Ann. Entomol. Soc. Am. 80: 212–216.

Hollis, P. L. 2000. U. S. insect losses variable during 1999. Southeast Farm Press 27(5): 35, 43.

Horton, D. L., J. D. Dutcher, H C Ellis, J. A. Payne, and C. E. Yonce. 1993. Peach insects, pp. 26–27. *In* R. M. McPherson and G. K. Douce (Eds.), Summary of losses from insect damage and costs of control in Georgia, 1992. Georgia Agric. Exp. Stn. Spec. Publ. 83: 1–55.

Hudson, R. D., and J. N. All. 1993. Grain sorghum insects, p. 13. *In* R. M. McPherson and G. K. Douce (Eds.), Summary of losses from insect damage and costs of control in Georgia, 1992. Georgia Agric. Exp. Stn. Spec. Publ. 83: 1–55.

Hudson, R. D., and G. D. Buntin. 1993. Small grain insects, p. 36. *In* R. M. McPherson and G. K. Douce (Eds.), Summary of losses from insect damage and costs of control in Georgia, 1992. Georgia Agric. Exp. Stn. Spec. Publ. 83: 1–55.

Hudson, R. D., D. C. Jones, and R. M. McPherson. 1993. Soybean insects, pp. 37–38. *In* R. M. McPherson and G. K. Douce (Eds.), Summary of losses from insect damage and costs of control in Georgia, 1992. Georgia Agric. Exp. Stn. Spec. Publ. 83: 1–55.

Hunter, R. E., and T. F. Leigh. 1965. A laboratory life history of the consperse stink bug, *Euschistus conspersus* (Hemiptera: Pentatomidae). Ann. Entomol. Soc. Am. 58: 648–649.

Ingram, J. W. 1927. Insects injurious to the rice crop. U.S.D.A. Farmers' Bull. 1543: 1–16.

International Rice Research Institute. 1997a. Importance of rice, pp. 7–9. *In* International Rice Research Institute Rice Almanac (2nd ed.), Manila, Philippines. 181 pp.

International Rice Research Institute. 1997b. Rice around the world, pp. 29–142. *In* Rice Research Institute Rice Almanac (2nd ed.), Manila, Philippines. 181 pp.

Ishiwatari, T. 1974. Studies on the scent of stink bugs (Hemiptera: Pentatomidae). I. Alarm pheromone activity. Appl. Entomol. Zool. 9: 153–158.

Ishiwatari, T. 1976. Studies on the scent of stink bugs (Hemiptera: Pentatomidae). II. Aggregation pheromone activity. Appl. Entomol. Zool. 11: 38–44.

Ito, K. 1978. Ecology of the stink bugs causing pecky rice. Rev. Plant Prot. Res. 11: 62–78.

Jacobson, L. A. 1936. Say's grain bug, *Chlorochroa sayi* Stal, in Canada. Can. Entomol. 68: 259–260.

Jacobson, L. A. 1940. Say's grain bug in western Canada. Cargill Crop. Bull. 15(17): 35–38.

Jacobson, L. A. 1945. The effect of Say stinkbug feeding on wheat. Can. Entomol. 77: 200.

Jacobson, L. A. 1965. Damage to wheat by Say stink bug, *Chlorochroa sayi*. Can. J. Plant Sci. 45: 413–417.

Javahery, M. 1990. Biology and ecological adaptation of the green stink bug (Hemiptera: Pentatomidae) in Quebec and Ontario. Ann. Entomol. Soc. Am. 83: 201–206.

Javahery, M. 1994. Development of eggs in some true bugs (Hemiptera-Heteroptera). Part I. Pentatomoidea. Can. Entomol. 126: 401–433.

Jensen, R. L., and J. Gibbens. 1973. Rearing the southern green stink bug on an artificial diet. J. Econ. Entomol. 66: 269–271.

Jensen, R. L., and L. D. Newsom. 1972. Effect of stink bug-damaged soybean seeds on germination, emergence, and yield. J. Econ. Entomol. 65: 261–264.

Jewett, H. H. 1959. Controlling tobacco insects. Kentucky Agric. Ext. Serv. Circ. 525: 1–38.

Johnson, D. R., J. J. Kimbrough, and M. L. Wall. 1987. Control of insects attacking rice. Arkansas Coop. Ext. Serv. EL 330: 1–15. (revised).

Jones, D. B., and R. H. Cherry. 1984. 1983 Stink bug studies. Florida Agric. Coop. Ext. Serv., Belle Glade Everglades Res. Educ. Center, Res. Rep. EV-1984–10: 17–21.

Jones, D. B., and R. H. Cherry. 1986. Species composition and seasonal abundance of stink bugs (Heteroptera: Pentatomidae) in southern Florida rice. J. Econ. Entomol. 79: 1226–1229.

Jones, T. H. 1918. The southern green plant-bug. U.S.D.A. Bull. 689: 1–27.

Jones, S. B., Jr., and A. E. Luchsinger. 1986. Plant systematics. McGraw-Hill, Inc., New York, NY. 512 pp.

Jones, V. P., and L. C. Caprio. 1992. Damage estimates and population trends of insects attacking seven macadamia cultivars in Hawaii. J. Econ. Entomol. 85: 1884–1890.

Jones, V. P., and L. C. Caprio. 1994. Southern green stink bug (Hemiptera: Pentatomidae) feeding on Hawaiian macadamia nuts: the relative importance of damage occurring in the canopy and on the ground. J. Econ. Entomol. 87: 431–435.

Jones, W. A., Jr. 1985. *Nezara viridula*, pp. 339–343. *In* P. Singh and R. F. Moore (Eds.), Handbook of insect rearing Vol. 1. Elsevier, New York, NY. 488 pp.

Jones, W. A. 1988. World review of the parasitoids of the southern green stink bug, *Nezara viridula* (L.) (Heteroptera: Pentatomidae). Ann. Entomol. Soc. Am. 81: 262–273.

Jones, W. A. 1993. New host and habitat associations for some Arizona Pentatomoidea and Coreidae. Southwestern Entomol. Suppl. 16: 1–29.

Jones, W. A., Jr., and F. D. Brewer. 1987. Suitability of various host plant seeds and artificial diets for rearing *Nezara viridula* (L.). J. Agric. Entomol. 4: 223–232.

Jones, W. A., Jr., and M. J. Sullivan. 1978. Susceptibility of certain soybean cultivars to damage by stink bugs. J. Econ. Entomol. 71: 534–536.

Jones, W. A., Jr., and M. J. Sullivan. 1979. Soybean resistance to the southern green stink bug, *Nezara viridula*. J. Econ. Entomol. 72: 628–632.

Jones, W. A., Jr., and M. J. Sullivan. 1981. Overwintering habitats, spring emergence patterns, and winter mortality of some South Carolina Hemiptera. Environ. Entomol. 10: 409–414.

Jones, W. A., and M. J. Sullivan. 1982. Role of host plants in population dynamics of stink bug pests of soybean in South Carolina. Environ. Entomol. 11: 867–875.

Jones, W. A., Jr., and M. J. Sullivan. 1983. Seasonal abundance and relative importance of stink bugs in soybean. South Carolina Agric. Exp. Stn. Tech Bull. 1087. 6 pp.

Jones, W. A., Jr., S. Y. Young, M. Shepard, and W. H. Whitcomb. 1983. Use of imported natural enemies against insect pests of soybean, pp. 63–77. *In* H. Pitre (Ed.), Natural enemies of arthropod pests in soybean. South Carolina Agric. Exp. Stn. (Southern Coop. Ser.) Bull. 285: 1–90.

Jones, W. A., B. M. Shepard, and M. J. Sullivan. 1996. Incidence of parasitism of pentatomid (Heteroptera) pests of soybean in South Carolina with a review of studies in other states. J. Agric. Entomol. 13: 243–263.

Kariya, H. 1961. Effect of temperature on the development and the mortality of the southern green stink bug, *Nezara viridula* and the oriental green stink bug, *N. antennata*. Japanese J. Appl. Entomol. and Zool. 5: 191–196.

Kelsheimer, E. G., and D. O. Wolfenbarger. 1952. Insects of tomatoes and their control. Florida Agric. Exp. Stn. Circ. S-51: 1–11.

Kester, K. M., C. M. Smith, and D. F. Gilman. 1984. Mechanisms of resistance in soybean (*Glycine max* [L.] Merrill) genotype PII71444 to the southern green stink bug, *Nezara viridula* (L.) (Hemiptera: Pentatomidae). Environ. Entomol. 13: 1208–1215.

Khan, M. F. R., and R. P. Griffin. 1999. Efficacy of insecticides on tomato insect pests, 1998. Arthropod Manage. Tests 24: 178–179.

Kilpatrick, R. A., and E. E. Hartwig. 1955. Fungus infection of soybean seed as influenced by stink bug injury. Plant Dis. Rep. 39(2): 177–180.

Kiritani, K. 1963. The change in reproductive system of the southern green stink bug, *Nezara viridula*, and its application to forecasting of the seasonal history. Japanese J. Appl. Entomol. Zool. 7: 327–337.

Kiritani, K. 1964a. The effect of colony size upon the survival of larvae of the southern green stink bug, *Nezara viridula*. Japanese J. Appl. Entomol. Zool. 8: 45–54.

Kiritani, K. 1964b. Natural control of populations of the southern green stink bug, *Nezara viridula*. Res. Pop. Ecol. 6: 88–98.

Kiritani, K. 1965. The natural regulation of the population of the southern green stink bug, *Nezara viridula* L. 12th Internatl. Cong. Entomol. (Section 6: Ecology) Proc.: 375.

Kiritani, K. 1970. Studies on the adult polymorphism in the southern green stink bug, *Nezara viridula* (Hemiptera: Pentatomidae). Res. Pop. Ecol. 12: 19–34.

Kiritani, K., and N. Hokyo. 1962. Studies on the life table of the southern green stink bug, *Nezara viridula*. Japanese J. Appl. Entomol. Zool. 6: 124–140.

Kiritani, K., and N. Hokyo. 1965. Variation of egg mass size in relation to the oviposition pattern in Pentatomidae. Kontyu 33: 427–433.

Kiritani, K., and K. Kimura. 1965. The effect of population density during nymphal and adult stages on the fecundity and other reproductive performances. Japanese J. Ecol. 15: 233–236.

Kiritani, K., N. Hokyo, and K. Kimura. 1963. Survival rate and reproductivity of the adult southern green stink bug, *Nezara viridula*, in the field cage. Japanese J. Appl. Entomol. Zool. 7: 113–124.

Kiritani, K., N. Hokyo, K. Kimura, and F. Nakasuji. 1965. Imaginal dispersal of the southern green stink bug, *Nezara viridula* L., in relation to feeding and oviposition. Japanese J. Appl. Entomol. Zool. 9: 291–297.

Kiritani, K., N. Hokyo, and K. Kimura. 1966a. Factors affecting the winter mortality in the southern green stink bug, *Nezara viridula* L. Ann. Soc. Entomol. France (N.S.) 2: 199–207.

Kiritani, K., F. Nakasuji, and N. Hokyo. 1966b. The survival rate of eggs and larvae in relation to group size in the southern green stink bug, *Nezara viridula* L. Japanese J. Appl. Entomol. Zool. 10: 205–211.

Kiritani, K., N. Hokyo, and S. Iwao. 1966c. Population behavior of the southern green stink bug, *Nezara viridula*, with special reference to the developmental stages of early-planted paddy. Res. Pop. Ecol. 8: 133–146.

Knowlton, G. F. 1944. Pentatomidae eaten by Utah birds. J. Econ. Entomol. 37: 118–119.

Knowlton, G. F. 1953. Say's stinkbug in Utah. Utah Agric. Exp. Stn., Mimeogr. Ser. 408: 1–8.

Kogan, M., and S. G. Turnipseed. 1987. Ecology and management of soybean arthropods. Annu. Rev. Entomol. 32: 507–538.

Krispyn, J. W., and J. W. Todd. 1982. The red imported fire ant as a predator of the southern green stink bug on soybean in Georgia. J. Georgia Entomol. Soc. 17: 19–26.

Krombein, K. V., P. D. Hurd, Jr., D. R. Smith, and B. D. Burks. 1979a. Catalog of Hymenoptera in America north of Mexico. Volume 1. Symphyta and Apocrita (Parasitica). XVI + pp. 1–1198. Smithsonian Inst. Press, Washington, D. C.

Krombein, K. V., P. D. Hurd, Jr., D. R. Smith, and B. D. Burks. 1979b. Catalog of Hymenoptera in America north of Mexico. Volume 2. Apocrita (Aculeata). XVI + pp. 1199–2209. Smithsonian Inst. Press, Washington, D. C.

Krombein, K. V., P. D. Hurd, Jr., and D. R. Smith. 1979c. Catalog of Hymenoptera in America north of Mexico. Volume 3. Indexes. XXX + pp. 2211–2735. Smithsonian Inst. Press, Washington, D. C.

Lam, W. F., and L. P. Pedigo. 1998. Response of soybean insect communities to row width under crop-residue management systems. Environ. Entomol. 27: 1069–1079.

Lambert, W. R., and G. A. Herzog. 1993. Cotton insects, pp. 7–8. *In* R. M. McPherson and G. K. Douce (Eds.), Summary of losses from insect damage and costs of control in Georgia, 1992. Georgia Agric. Exp. Stn. Spec. Publ. 83: 1–55.

Lanigan, P. J., and E. J. Barrows. 1977. Sexual behavior of *Murgantia histrionica* (Hemiptera: Pentatomidae). Psyche 84: 191–197.

Leach, J. G., and G. Clulo. 1943. Association between *Nematospora phaseoli* and the green stinkbug. Phytopathology 33: 1209–1211.

Lee, F. N., and N. P. Tugwell. 1980. "Pecky rice" and quality reduction in Arkansas rice production. Arkansas Farm Res. 29(5): 2.

Lee, F. N., N. P. Tugwell, G. J. Weidemann, and W. C. Smith. 1986. Microorganisms associated with pecky rice. Texas Agric. Exp. Stn., Proc. 21st Rice Tech. Working Group, p. 90.

Lee, F. N., N. P. Tugwell, S. J. Fannah, and G. J. Weidemann. 1993. Role of fungi vectored by rice stink bug (Heteroptera: Pentatomidae) in discoloration of rice kernels. J. Econ. Entomol. 86: 549–556.

Leiby, R. W. 1928. Cotton boll weevil damage during 1927. J. Econ. Entomol. 21: 151.

Lingren, P. S., A. T. Wier, and D. J. Boethel. 1995. Efficacy of selected insecticides against the southern green stink bug on soybean in Louisiana, 1994. Arthropod Manage. Tests 20: 236–237.

Lockwood, J. A., and R. N. Story. 1985a. Bifunctional pheromone in the first instar of the southern green stink bug, *Nezara viridula* (L.) (Hemiptera: Pentatomidae): its characterization and interaction with other stimuli. Ann. Entomol. Soc. Am. 78: 474–479.

Lockwood, J. A., and R. N. Story. 1985b. Photic, thermic, and sibling influences on the hatching rhythm of the southern green stink bug, *Nezara viridula* (L.). Environ. Entomol. 14: 562–567.

Lockwood, J. A., and R. N. Story. 1985c. The diurnal ethology of the adult green stink bug, *Acrosternum hilare*, in senescing soybeans. J. Entomol. Sci. 20: 69–75.

Lockwood, J. A., and R. N. Story. 1986a. Adaptive functions of nymphal aggregation in the southern green stink bug, *Nezara viridula* (L.) (Hemiptera: Pentatomidae). Environ. Entomol. 15: 739–749.

Lockwood, J. A., and R. N. Story. 1986b. The diurnal ethology of the southern green stink bug, *Nezara viridula* (L.) in cowpeas. J. Entomol. Sci. 21: 175–184.

Lockwood, J. A., and R. N. Story. 1986c. Embryonic orientation in pentatomids: its mechanism and function in southern green stink bug (Hemiptera: Pentatomidae). Ann. Entomol. Soc. Am. 79: 963–970.

Lockwood, J. A., and R. N. Story. 1987. Defensive secretion of the southern green stink bug (Hemiptera: Pentatomidae) as an alarm pheromone. Ann. Entomol. Soc. Am. 80: 686–691.

Luck, R. F. 1981. Parasitic insects introduced as biological control agents for arthropod pests, pp. 125–284. *In* D. Pimentel (Ed.), CRC handbook of pest management in agriculture, volume II. CRC Press, Inc., Boca Raton, FL. 501 pp.

Ludwig, S. W., and L. T. Kok. 1998a. Phenology and parasitism of harlequin bugs, *Murgantia histrionica* (Hahn) (Hemiptera: Pentatomidae), in southwest Virginia. J. Entomol. Sci. 33: 33–39.

Ludwig, S. W., and L. T. Kok. 1998b. Evaluation of trap crops to manage harlequin bugs, *Murgantia histrionica* (Hahn) (Hemiptera: Pentatomidae) on broccoli. Crop Prot. 17: 123–128.

Lugger, O. 1900. Bugs injurious to our cultivated plants. Minnesota Agric. Exp. Stn. Bull. 69: 1–259.

Lye, B.-H., and R. N. Story. 1988. Feeding preference of the southern green stink bug (Hemiptera: Pentatomidae) on tomato fruit. J. Econ. Entomol. 81: 522–526.

Lye, B.-H., and R. N. Story. 1989. Spatial dispersion and sequential sampling plan of the southern green stink bug (Hemiptera: Pentatomidae) on fresh market tomatoes. Environ. Entomol. 18: 139–144.

Lye, B.-H., R. N. Story, and V. L. Wright. 1988a. Southern green stink bug (Hemiptera: Pentatomidae) damage to fresh market tomatoes. J. Econ. Entomol. 81: 189–194.

Lye, B.-H., R. N. Story, and V. L. Wright. 1988b. Damage threshold of the southern green stink bug, *Nezara viridula*, (Hemiptera: Pentatomidae) on fresh market tomatoes. J. Entomol. Sci. 23: 366–373.

MacNay, C. G. 1952. Summary of the more important insect infestations and occurrences in Canada in 1951. Annu. Rep. Entomol. Soc. Ontario 82: 91–115.

MacNay, C. G. 1958. Summary of important insect infestations, occurrences and damage in Canada in 1957. Annu. Rep. Entomol. Soc. Ontario 88: 63–78.

Mahaffey, J. S., J. S. Bacheler, J. R. Bradley, Jr., and J. W. Van Duyn. 1994. Performance of Monsanto's transgenic *B. t.* cotton against high populations of lepidopterous pests in North Carolina. Proc. 1994 Beltwide Cotton Conf. 2: 1061–1063.

Marchetti, M. A., and H. D. Petersen. 1984. The role of *Bipolaris oryzae* in floral abortion and kernel discoloration in rice. Plant Dis. 68: 288–291.

Marsolan, N. F., and W. G. Rudd. 1976. Modeling and optimal control of insect pest populations. Math. Biosci. 30: 231–244.

Mattiacci, L., S. B. Vinson, H. J. Williams, J. R. Aldrich, and F. Bin. 1993. A long-range attractant kairomone for egg parasitoid *Trissolcus basalis*, isolated from defensive secretion of its host, *Nezara viridula*. J. Chem. Ecol. 19: 1167–1181.

Mau, R., W. C. Mitchell, and M. Anwar. 1967. Preliminary studies on the effects of gamma irradiation of eggs and adults of the southern green stink bug, *Nezara viridula* (L.). Proc. Hawaiian Entomol. Soc. 19: 415–417.

Mauney, J. R. 1986. Vegetative growth and development of fruiting sites, pp. 11–28. *In* J. R. Mauney and J. McD. Stewart (Eds.), Cotton physiology. Cotton Found., Reference Book Ser., Memphis, TN. 786 pp.

McLain, D. K. 1980. Female choice and the adaptive significance of prolonged copulation in *Nezara viridula* (Hemiptera: Pentatomidae). Psyche 87: 325–336.

McLain, D. K. 1981a. Sperm precedence and prolonged copulation in the southern green stink bug, *Nezara viridula*. J. Georgia Entomol. Soc. 16: 70–77.

McLain, D. K. 1981b. Numerical responses of *Murgantia histrionica* to concentrations of its host plant. J. Georgia Entomol. Soc. 16: 257–260.

McLain, D. K. 1985. Male size, sperm competition, and the intensity of sexual selection in the southern green stink bug, *Nezara viridula* (Hemiptera: Pentatomidae). Ann. Entomol. Soc. Am. 78: 86–89.

McLain, D. K. 1987. Heritability of size, a sexually selected character, and the response to sexual selection in a natural population of the southern green stink bug, *Nezara viridula* (Hemiptera: Pentatomidae). Heredity 59: 391–395.

McLain, D. K. 1991. Heritability of size: a positive correlate of multiple fitness components in the southern green stink bug (Hemiptera: Pentatomidae). Ann. Entomol. Soc. Am. 84: 174–178.

McLain, D. K. 1992. Preference for polyandry in female stink bugs, *Nezara viridula* (Hemiptera: Pentatomidae). J. Insect Behav. 5: 403–410.

McLain, D. K., and N. B. Marsh. 1990a. Male copulatory success: heritability and relationship to mate fecundity in the southern green stinkbug, *Nezara viridula* (Hemiptera: Pentatomidae). Heredity 64: 161–167.

McLain, D. K., and N. B. Marsh. 1990b. Individual sex ratio adjustment in response to the operational sex ratio in the southern green stinkbug. Evolution 44: 1018–1025.

McLain, D. K., N. B. Marsh, J. R. Lopez, and J. A. Drawdy. 1990a. Intravernal changes in the level of parasitization of the southern green stink bug (Hemiptera: Pentatomidae), by the feather-legged fly (Diptera: Tachinidae): host sex, mating status, and body size as correlated factors. J. Entomol. Sci. 25: 501–509.

McLain, D. K., D. L. Lanier, and N. B. Marsh. 1990b. Effects of female size, mate size, and number of copulations on fecundity, fertility, and longevity of *Nezara viridula* (Hemiptera: Pentatomidae). Ann. Entomol. Soc. Am. 83: 1130–1136.

McMurran, S. M., and J. B. Demaree. 1920. Diseases of southern pecans. U.S.D.A. Farmers' Bull. 1129: 1–22.

McPherson, J. E. 1971. Laboratory rearing of *Euschistus tristigmus tristigmus*. J. Econ. Entomol. 64: 1339–1340.

McPherson, J. E. 1975. Life history of *Euschistus tristigmus tristigmus* (Hemiptera: Pentatomidae) with information on adult seasonal dimorphism. Ann. Entomol. Soc. Am. 68: 333–334.

McPherson, J. E. 1982. The Pentatomoidea (Hemiptera) of northeastern North America with emphasis on the fauna of Illinois. Southern Illinois Univ. Press, Carbondale and Edwardsville. 240 pp.

McPherson, J. E., and J. P. Cuda. 1974. The first record in Illinois of *Nezara viridula* (Hemiptera: Pentatomidae). Trans. Illinois State Acad. Sci. 67: 461–462.

McPherson, J. E., and R. H. Mohlenbrock. 1976. A list of the Scutelleroidea of the La Rue-Pine Hills Ecological Area with notes on biology. Great Lakes Entomol. 9: 125–169.

McPherson, J. E., and D. L. Tecic. 1997. Notes on the life histories of *Acrosternum hilare* and *Cosmopepla bimaculata* (Heteroptera: Pentatomidae) in southern Illinois. Great Lakes Entomol. 30: 79–84.

McPherson, R. M. 1996. Relationship between soybean maturity group and the phenology and abundance of stink bugs (Heteroptera: Pentatomidae): impact on yield and quality. J. Entomol. Sci. 31: 199–208.

McPherson, R. M., and K. Bondari. 1991. Influence of planting date and row width on abundance of velvetbean caterpillars (Lepidoptera: Noctuidae) and southern green stink bugs (Heteroptera: Pentatomidae) in soybean. J. Econ. Entomol. 84: 311–316.

McPherson, R. M., and G. K. Douce. 1993. List of the 20 most damaging insect species or complexes in Georgia in 1992, p. 48. *In* R. M. McPherson and D. C. Douce (Eds.), Summary of losses from insect damage and costs of control in Georgia, 1992. Georgia Agric. Exp. Stn. Spec. Publ. 83: 1–55.

McPherson, R. M., and D. C. Jones. 1993. Tobacco insects, pp. 40–41. *In* R. M. McPherson and G. K Douce (Eds.), Summary of losses from insect damage and costs of control in Georgia, 1992. Georgia Agric. Exp. Stn. Spec. Publ. 83: 1–55.

McPherson, R. M., and L. D. Newsom. 1984. Trap crops for control of stink bugs in soybean. J. Georgia Entomol. Soc. 19: 470–480.

McPherson, R. M., L. D. Newsom, and B. F. Farthing. 1979a. Evaluation of four stink bug species from three genera affecting soybean yield and quality in Louisiana. J. Econ. Entomol. 72: 188–194.

McPherson, R. M., J. B. Graves, and T. A. Allain. 1979b. Dosage-mortality responses and field control of seven pentatomids, associated with soybean, exposed to methyl parathion. Environ. Entomol. 8: 1041–1043.

McPherson, R. M., J. R. Pitts, L. D. Newsom, J. B. Chapin, and D. C. Herzog. 1982. Incidence of tachinid parasitism of several stink bug (Heteroptera: Pentatomidae) species associated with soybean. J. Econ. Entomol. 75: 783–786.

McPherson, R. M., G. W. Zehnder, and J. C. Smith. 1988. Influence of cultivar, planting date, and row width on abundance of green cloverworms (Lepidoptera: Noctuidae) and green stink bugs (Heteroptera: Pentatomidae) in soybean. J. Entomol. Sci. 23: 305–313.

McPherson, R. M., G. K. Douce, and R. D. Hudson. 1993. Annual variation in stink bug (Heteroptera: Pentatomidae) seasonal abundance and species composition in Georgia soybean and its impact on yield and quality. J. Entomol. Sci. 28: 61–72.

McPherson, R. M., J. W. Todd, and K. V. Yeargan. 1994. Stink bugs, pp. 87–90. *In* L. G. Higley and D. J. Boethel (Eds.), Handbook of soybean insect pests. Entomol. Soc. Am. Publ., Lanham, MD. 136 pp.

McPherson, R. M., R. D. Hudson, and D. C. Jones. 1995a. Soybean insects, pp. 38–39. *In* G. K. Douce and R. M. McPherson (Eds.), Summary of losses from insect damage and costs of control in Georgia 1993. Georgia Agric. Exp. Stn. Spec. Publ. 87: 1–56.

McPherson, R. M., D. J. Boethel, J. E. Funderburk, and A. T. Wier. 1995b. The effect of alternative southern green stink bug (Heteroptera: Pentatomidae) insecticide controls on soybean pest management, quality and yield. J. Entomol. Sci. 30: 216–236.

McPherson, R. M., T. P. Mack, J. E. Funderburk, D. J. Boethel, C. G. Helm, M. Kogan, L. Lambert, G. L. Lentz, H. N. Pitre, M. J. Sullivan, and M. O. Way. 1996. A multi-state field evaluation of arthropod pest resistance in soybean and impact on natural enemies. Georgia Agric. Exp. Stn. Spec. Publ. 88: 1–77.

McPherson, R. M., R. C. Layton, W. J. McLaurin, and W. A. Mills, III. 1998a. Influence of irrigation and maturity group on the seasonal abundance of soybean arthropods. J. Entomol. Sci. 33: 378–392.

McPherson, R. M., J. D. Taylor, and B. D. Crowe. 1998b. Control of insect pests on Georgia soybeans, 1997. Arthropod Manage. Tests 23: 283–284.

McPherson, R. M., D. Boethel, and M. E. Baur. 1999a. Adoption of early soybean production systems in the southern U.S.: impact on soybean IPM system, pp. 320–330. *In* H. E. Kauffman (Ed.), World Soybean Research Conference VI: Proceedings. Superior Printing, Champaign, IL. 746 pp.

McPherson, R. M., M. L. Wells, and C. S. Bundy. 1999b. Control of stink bugs and velvetbean caterpillars on Georgia soybeans, 1998. Arthropod Manage. Tests 24: 289–290.

McPherson, R. M., J. D. Taylor, and B. D. Crowe. 1999c. Control of insect pests on soybeans, 1998. Arthropod Manage. Tests 24: 291.

McPherson, R. M., M. L. Wells, and C. S. Bundy. 2000. Impact of the early soybean production system on arthropod pest populations in Georgia. Environ. Entomol. (in press).

Mead, F. W. 1983. Insect detection: a stink bug, *Oebalus grisescens* (Sailer). Tri-ology 22(11): 4.

Menezes, E. B., D. C. Herzog, I. D. Teare, and R. K. Sprenkel. 1985a. Phenological events affecting southern green stink bug on soybean. Soil and Crop Sci. Soc. Florida Proc. 44: 227–231.

Menezes, E. B., D. C. Herzog, and P. J. d'Almada. 1985b. A study of parasitism of the southern green stink bug, *Nezara viridula* (L.) (Hemiptera: Pentatomidae), by *Trichopoda pennipes* (F.) (Diptera: Tachinidae). Ann. Soc. Entomol. Bras. 14: 29–35.

Menusan, H., Jr. 1943. Plant bugs, pp. 29–30. *In* F. L. Campbell and F. R. Moulton (Eds.), Laboratory procedures in studies of the chemical control of insects. Am. Assoc. Adv. Sci. Publ. 20: 1–206.

Merchant, M. E., and G. L. Teetes. 1992. Evaluation of selected sampling methods for panicle-infesting insect pests of sorghum. J. Econ. Entomol. 85: 2418–2424.

Michailides, T. J., R. E. Rice, and J. M. Ogawa. 1987. Succession and significance of several hemipterans attacking a pistachio orchard. J. Econ. Entomol. 80: 398–406.

Michailides, T. J., J. M. Ogawa, and R. E. Rice. 1988. Sites of epicarp lesion and kernel necrosis in relationship to symptoms and phenology of pistachio fruit. J. Econ. Entomol. 81: 1152–1154.

Michelbacher, A. E., W. W. Middlekauff, and O. G. Bacon. 1952. Stink bug injury to tomatoes in California. J. Econ. Entomol. 45: 126.

Miles, P. W. 1968. Insect secretions in plants. Ann. Rev. Phytopathol. 6: 137–164.

Millar, J. G. 1997. Methyl (2E,4Z,6Z)-deca-2,4,6–trienoate, a thermally unstable, sex-specific compound from the stink bug *Thyanta pallidovirens*. Tetrahedron Letters 38: 7971–7972.

Miner, F. D. 1961. Stink bug damage to soybeans. Arkansas Agric. Exp. Stn., Arkansas Farm Res. 10(3): 12.

Miner, F. D. 1966a. Biology and control of stink bugs on soybeans. Arkansas Agric. Exp. Stn. Bull. 708: 1–40.

Miner, F. D. 1966b. The stink bug problem on soybeans. Soybean Digest 26: 14–16.

Miner, F. D., and B. Dumas. 1980a. Stored soybeans and stink bug damage. Arkansas Agric. Exp. Stn., Arkansas Farm Res. 29(4): 14.

Miner, F. D., and B. A. Dumas. 1980b. Effect of green stink bug damage on soybean seed quality before and after storage. Arkansas Agric. Exp. Stn. Bull. 844: 1–19.

Miner, F. D., and T. H. Wilson. 1966. Quality of stored soybeans as affected by stink bug damage. Arkansas Agric. Exp. Stn., Arkansas Farm Res. 15(6): 2.

Mitchell, W. C. 1965. An example of integrated control of insects: status of the southern green stink bug in Hawaii. Agric. Sci. Rev. 3: 32–35.

Mitchell, W. C., and R. F. L. Mau. 1969. Sexual activity and longevity of the southern green stink bug, *Nezara viridula*. Ann. Entomol. Soc. Am. 62: 1246–1247.

Mitchell, W. C., and R. F. L. Mau. 1971. Response of the female southern green stink bug and its parasite, *Trichopoda pennipes*, to male stink bug pheromones. J. Econ. Entomol. 64: 856–859.

Mitchell, W. C., R. M. Warner, and E. T. Fukunaga. 1965. Southern green stink bug, *Nezara viridula* (L.), injury to macadamia nut. Proc. Hawaiian Entomol. Soc. 19: 103–109.

Mizell, R. F., III, and W. L. Tedders. 1995. A new monitoring method for detection of the stinkbug complex in pecan orchards. Proc. Southeastern Pecan Growers Assoc. 88: 36–40.

Mizell, R. F., III, H C Ellis, and W. L. Tedders. 1996. Traps to monitor stink bugs and pecan weevils. The Pecan Grower 8(2): 17–20.

Mizell, R. F., III, W. L. Tedders, C. E. Yonce, and J. A. Aldrich. 1997. Stink bug monitoring — an update. Proc. Southeastern Pecan Growers Assoc. 90: 50–52.

Morrill, A. W. 1905. Report on a Mexican cotton pest, the "conchuela." U.S.D.A. Bur. Entomol. Bull. 54: 18–34.

Morrill, A. W. 1907. The Mexican conchuela in western Texas in 1905. U.S.D.A. Bur. Entomol. Bull. 64 (Part I): 1–14.

Morrill, A. W. 1910. Plant-bugs injurious to cotton bolls. U.S.D.A. Bur. Entomol. Bull. (N.S.) 86: 1–110.

Morrill, A. W. 1912. Report of the entomologist of the Arizona Horticultural Commission for the year ending June 30, 1912, pp. 15–43. *In* 4th Annu. Rep. Arizona Hortic. Comm. 1912: 1–43.

Moznette, G. F., C. B. Nickels, W. C. Pierce, T. L. Bissell, J. B. Demaree, J. R. Cole, H. E. Parson, and J. R. Large. 1940. Insects and diseases of the pecan and their control. U.S.D.A. Farmers' Bull. 1829: 1–70.

Mundinger, F. G. 1940. Pentatomids attacking tomatoes and experiments on their control. J. Econ. Entomol. 33: 275–278.

Mundinger, F. G., and P. J. Chapman. 1932. Plant bugs as pests of pear and other fruits in the Hudson Valley. J. Econ. Entomol. 25: 655–658.

Munro, J. A., and F. G. Butcher. 1940. Say's stink bug. North Dakota Agric. Exp. Stn. Bimonthly Bull. 3(2): 11–13.

Munyaneza, J., and J. E. McPherson. 1994. Comparative study of life histories, laboratory rearing, and immature stages of *Euschistus servus* and *Euschistus variolarius* (Hemiptera: Pentatomidae). Great Lakes Entomol. 26: 263–274.

Naresh, J. S., and C. M. Smith. 1983. Development and survival of rice stink bugs (Hemiptera: Pentatomidae) reared on different host plants at four temperatures. Environ. Entomol. 12: 1496–1499.

Naresh, J. S., and C. M. Smith. 1984. Feeding preference of the rice stink bug on annual grasses and sedges. Entomol. Exp. Appl. 35: 89–92.

Negron, J. F., and T. J. Riley. 1987. Southern green stink bug, *Nezara viridula* (Heteroptera: Pentatomidae), feeding in corn. J. Econ. Entomol. 80: 666–669.

Nettles, W. C., C. A. Thomas, and F. H. Smith. 1970. Soybean insects and diseases. How to control. South Carolina Ext. Serv. Circ. 504: 1–25.

Newsom, L. D., and D. C. Herzog. 1977. Trap crops for control of soybean pests. Louisiana Agric. 20(3): 14–15.

Newsom, L. D., M. Kogan, F. D. Miner, R. L. Rabb, S. G. Turnipseed, and W. H. Whitcomb. 1980. General accomplishments toward better pest control in soybean, pp. 51–98. *In* C. B. Huffaker (Ed.), New technology of pest control. Wiley-Interscience Publ., John Wiley & Sons, Inc., New York, NY. 500 pp.

Nilakhe, S. S. 1976a. Overwintering, survival, fecundity, and mating behavior of the rice stink bug. Ann. Entomol. Soc. Am. 69: 717–720.

Nilakhe, S. S. 1976b. Rice lines screened for resistance to the rice stink bug. J. Econ. Entomol. 69: 703–705.

Nilakhe, S. S., and R. B. Chalfant. 1982. Cowpea cultivars screened for resistance to insect pests. J. Econ. Entomol. 75: 223–227.

Nilakhe, S. S., R. B. Chalfant, and S. V. Singh. 1981a. Damage to southern peas by different stages of the southern green stink bug. J. Georgia Entomol. Soc. 16: 409–414.

Nilakhe, S. S., R. B. Chalfant, and S. V. Singh. 1981b. Evaluation of southern green stink bug damage to cowpeas. J. Econ. Entomol. 74: 589–592.

Nilakhe, S. S., R. B. Chalfant, and S. V. Singh. 1981c. Field damage to lima beans by different stages of southern green stink bug. J. Georgia Entomol. Soc. 16: 392–396.

Nishida, T. 1966. Behavior and mortality of the southern stink bug *Nezara viridula* in Hawaii. Res. Pop. Ecol. 8: 78–88.

Odglen, G. 1960. How much damage does the rice stink bug cause? Arkansas Farm Res. 9(1): 12.

Odglen, G. E., and L. O. Warren. 1962. The rice stink bug, *Oebalus pugnax* F. (sic), in Arkansas. Arkansas Agric. Exp. Stn. Rep. Ser. 107: 1–23.

Oetting, R. D., and T. R. Yonke. 1971. Biology of some Missouri stink bugs. J. Kansas Entomol. Soc. 44: 446–459.

Ohlendorf, B. L. P. 1996. Integrated pest management for cotton in the western region of the United States (2nd ed.). Univ. California, Div. Agric. Nat. Res. 3305: 1–164.

Oi, D. H. 1991. Investing in biological control: initiation of a parasite mass rearing program for macadamia nut orchards in Hawaii. Hawaiian Coop. Ext. Serv., 1989 ADAP Crop Prot. Conf. Proc., M. W. Johnson, D. E. Ullman, and A. Vargo (Eds.). Hawaiian Res. Ext. Series 134: 128–130.

O'Keeffe, L. E., H. W. Homan, and D. J. Schotzko. 1991. Chalky spot damage to lentils. Idaho Agric. Exp. Stn., Current Inform. Ser. 894. 4 pp.

Oliver, B. F., J. R. Gifford, and G. B. Trahan. 1972. Evaluation of insecticidal sprays for controlling the rice stinkbug in southwest Louisiana. J. Econ. Entomol. 65: 268–270.

Olsen, C. E. 1912. Contribution to an annotated list of Long Island insects. J. New York Entomol. Soc. 20: 48–58.

Orr, D. B., D. J. Boethel, and W. A. Jones. 1985a. Development and emergence of *Telenomus chloropus* and *Trissolcus basalis* (Hymenoptera: Scelionidae) at various temperatures and relative humidities. Ann. Entomol. Soc. Am. 78: 615–619.

Orr, D. B., D. J. Boethel, and W. A. Jones. 1985b. Biology of *Telenomus chloropus* (Hymenoptera: Scelionidae) from eggs of *Nezara viridula* (Hemiptera: Pentatomidae) reared on resistant and susceptible soybean genotypes. Can. Entomol. 117: 1137–1142.

Orr, D. B., J. S. Russin, D. J. Boethel, and W. A. Jones. 1986. Stink bug (Hemiptera: Pentatomidae) egg parasitism in Louisiana soybeans. Environ. Entomol. 15: 1250–1254.

Orr, D. B., D. J. Boethel, and M. B. Layton. 1989. Effect of insecticide applications in soybeans on *Trissolcus basalis* (Hymenoptera: Scelionidae). J. Econ. Entomol. 82: 1078–1084.

Osborn, H. 1894. Notes on the distribution of Hemiptera. Proc. Iowa Acad. Sci. 1 (Part 4): 120–123.

Osburn, M. R., A. M. Phillips, and W. C. Pierce. 1954. Insects and diseases of the pecan and their control. U.S.D.A. Farmers' Bull. 1829: 1–56.

Ota, D., and A. Cokl. 1991. Mate location in the southern green stink bug, *Nezara viridula* (Heteroptera: Pentatomidae), mediated through substrate-borne signals on ivy. J. Insect Behav. 4: 441–447.

Paddock, F. B. 1915. The harlequin cabbage-bug. Texas Agric. Exp. Stn. Bull. 179: 1–9.

Paddock, F. B. 1918. Studies on the harlequin bug. Texas Agric. Exp. Stn. Bull. 227: 1–65.

Panizzi, A. R. 1985. Dynamics of phytophagous pentatomids associated with soybean in Brazil, pp. 674–680. *In* R. Shibles (Ed.), World Soybean Research Conference III: Proceedings. Westview Press, Boulder, CO. 1,262 pp.

Panizzi, A. R. 1997. Wild hosts of pentatomids: ecological significance and role in their pest status on crops. Ann. Rev. Entomol. 42: 99–122.

Panizzi, A. R. 2000. Suboptimal nutrition and feeding behavior of hemipterans on less preferred plant food sources. Anais Soc. Entomol. Brasil 29: 1–12.

Panizzi, A. R., and R. M. L. Alves. 1993. Performance of nymphs and adults of the southern green stink bug (Heteroptera: Pentatomidae) exposed to soybean pods at different phenological stages of development. J. Econ. Entomol. 86: 1088–1093.

Panizzi, A. R., and D. C. Herzog. 1984. Biology of *Thyanta perditor* (Hemiptera: Pentatomidae). Ann. Entomol. Soc. Am. 77: 646–650.

Panizzi, A. R., and E. Hirose. 1995a. Survival, reproduction, and starvation resistance of adult southern green stink bug (Heteroptera: Pentatomidae) reared on sesame or soybean. Ann. Entomol. Soc. Am. 88: 661–665.

Panizzi, A. R., and E. Hirose. 1995b. Seasonal body weight, lipid content, and impact of starvation and water stress on adult survivorship and longevity of *Nezara viridula* and *Euschistus heros*. Entomol. Exp. Appl. 76: 247–253.

Panizzi, A. R., and A. M. Meneguim. 1989. Performance of nymphal and adult *Nezara viridula* on selected alternate host plants. Entomol. Exp. Appl. 50: 215–223.

Panizzi, A. R., and S. I. Saraiva. 1993. Performance of nymphal and adult southern green stink bug on an overwintering host plant and impact of nymph to adult food-switch. Entomol. Exp. Appl. 68: 109–115.

Panizzi, A. R., and F. Slansky, Jr. 1985a. Review of phytophagous pentatomids (Hemiptera: Pentatomidae) associated with soybean in the Americas. Florida Entomol. 68: 184–214.

Panizzi, A. R., and F. Slansky, Jr. 1985b. New host plant records for the stink bug *Piezodorus guildinii* in Florida (Hemiptera: Pentatomidae). Florida Entomol. 68: 215–216.

Panizzi, A. R., and F. Slansky, Jr. 1985c. Legume host impact on performance of adult *Piezodorus guildinii* (Westwood) (Hemiptera: Pentatomidae). Environ. Entomol. 14: 237–242.

Panizzi, A. R., and F. Slansky, Jr. 1985d. *Piezodorus guildinii* (Hemiptera: Pentatomidae): an unusual host of the tachinid *Trichopoda pennipes*. Florida Entomol. 68: 485–486.

Panizzi, A. R., and F. Slansky, Jr. 1985e. New host plant record for the stink bug *Thyanta custator* (Hemiptera: Pentatomidae) in Florida. Florida Entomol. 68: 693–694.

Panizzi, A. R., and F. Slansky, Jr. 1991. Suitability of selected legumes and the effect of nymphal and adult nutrition in the southern green stink bug (Hemiptera: Heteroptera: Pentatomidae). J. Econ. Entomol. 84: 103–113.

Panizzi, A. R., M. H. M. Galileo, H. A. O. Gastal, J. F. F. Toledo, and C. H. Wild. 1980. Dispersal of *Nezara viridula* and *Piezodorus guildinii* nymphs in soybeans. Environ. Entomol. 9: 293–297.

Panizzi, A. R., C. C. Niva, and E. Hirose. 1995. Feeding preference by stink bugs (Heteroptera: Pentatomidae) for seeds within soybean pods. J. Entomol. Sci. 30: 333–341.

Panizzi, A. R., L. M. Vivan, B. S. Correa-Ferreira, and L. A. Foerster. 1996. Performance of southern green stink bug (Heteroptera: Pentatomidae) nymphs and adults on a novel food plant (Japanese privet) and other hosts. Ann. Entomol. Soc. Am. 89: 822–827.

Panizzi, A. R., J. E. McPherson, D. G. James, M. Javahery, and R. M. McPherson. 2000. Stink bugs (Pentatomidae), pp. 421–474. *In* C. W. Schaefer and A. R. Panizzi (Eds.), Heteroptera of economic importance. CRC Press LLC, Boca Raton, FL. 828 pp.

Pantoja, A. 1990. Lista preliminar de plagas del arroz en Colombia. Arroz en las Americas 11(1): 9.

Pantoja, A., E. Daza, C. Garcia, O. I. Mejia, and D. A. Rider. 1995. Relative abundance of stink bugs (Hemiptera: Pentatomidae) in southwestern Colombia rice fields. J. Entomol. Sci. 30: 463–467.

Parish, H. E. 1934. Biology of *Euschistus variolarius* P. de B. (Family Pentatomidae; Order Hemiptera). Ann. Entomol. Soc. Am. 27: 50–54.

Pathak, M. D. 1968. Ecology of common insect pests of rice. Annu. Rev. Entomol. 13: 257–294.

Patton, R. L., and G. A. Mail. 1935. The grain bug (*Chlorochroa sayii* [sic] Stal) in Montana. J. Econ. Entomol. 28: 906–913.

Pavis, C., C. Malosse, P. H. Ducrot, and C. Descoins. 1994. Dorsal abdominal glands in nymphs of southern green stink bug, *Nezara viridula* (L.) (Heteroptera: Pentatomidae): chemistry of secretions of five instars and role of (E)-4–oxo-2–decenal, compound specific to first instars. J. Chem. Ecol. 20: 2213–2227.

Payne, J. A., and J. M. Wells. 1984. Toxic penicillia isolated from lesions of kernel-spotted pecans. Environ. Entomol. 13: 1609–1612.

Payne, J. A., H. L. Malstrom, and G. E. KenKnight. 1979. Insect pests and diseases of the pecan. U.S.D.A. Science and Education Administration, Agric. Reviews and Manuals. ARM-S-5. 43 pp.

Perry, E. J. 1979. Stink bugs in the home garden. California Coop. Ext. Leaflet 21106. 2 pp.

Phillips, A. M., J. R. Large, and J. R. Cole. 1964. Insects and diseases of the pecan in Florida. Florida Agric. Exp. Stn. Bull. 619A: 1–87.

Phillips, J. H. H. 1951. An annotated list of Hemiptera inhabiting sour cherry orchards in the Niagara Peninsula, Ontario. Can. Entomol. 83: 194–205.

Phillips, K. A., and J. O. Howell. 1980. Biological studies on two *Euchistus* [sic] species in central Georgia apple orchards. J. Georgia Entomol. Soc. 15: 349–355.

Pletsch, D. J. 1943. The Say stinkbug on potatoes in Montana. J. Econ. Entomol. 36: 334–335.

Polles, S. G. 1977. Black pit and kernel spot of pecans: special emphasis on southern green stink bug. Proc. Southeastern Pecan Growers Assoc. 70: 47–52.

Polles, S. G. 1979. Black pit and kernel spot of pecans: special emphasis on southern green stinkbug. Proc. 14th Annu. Conf., Georgia Pecan Growers Assoc. 10: 29–37.

Polles, S. G., J. W. Todd, and E. K. Heaton. 1973. Some insects causing black pit and kernel spot of pecans. Proc. 8th Annu. Conf., Georgia Pecan Growers Assoc. 4: 32–34.

Porter, B. A., S. C. Chandler, and R. F. Sazama. 1928. Some causes of cat-facing in peaches. Illinois Nat. Hist. Surv. Bull. 17: 261–275.

Powell, J. E., M. Shepard, and M. J. Sullivan. 1981. Use of heating degree day and physiological day equations for predicting development of the parasitoid *Trissolcus basalis*. Environ. Entomol. 10: 1008–1011.

Powell, J. E., M. Shepard, and P. T. Holmes. 1983. A stochastic model of parasitism of the southern green stink bug by *Trissolcus basalis* (Wollaston). South Carolina Agric. Exp. Stn. Tech. Bull. 1088. 7 pp.

Rabb, R. L., F. E. Guthrie, H. E. Scott, and C. F. Smith. 1959. Tobacco insects of North Carolina and their natural enemies. North Carolina Agric. Exp. Stn. Bull. 394: 1–32.

Race, S. R. 1960. A comparison of two sampling techniques for lygus bugs and stink bugs on cotton. J. Econ. Entomol. 53: 689–690

Radcliffe, E. B., K. L. Flanders, D. W. Ragsdale, and D. M. Noetzel. 1991. Pest management systems for potato insects, pp. 587–621. *In* D. Pimentel (Ed.), CRC handbook of pest management in agriculture (2nd ed.) Volume III. CRC Press, Inc., Boca Raton, FL. 749 pp.

Radford, A. E., H. E. Ahles, and C. R. Bell. 1981. Manual of the vascular flora of the Carolinas. Univ. North Carolina Press, Chapel Hill. 1,183 pp.

Ragsdale, D. W., A. D. Larson, and L. D. Newsom. 1979. Microorganisms associated with feeding and from various organs of *Nezara viridula*. J. Econ. Entomol. 72: 725–727.

Ragsdale, D. W., A. D. Larson, and L. D. Newsom. 1981. Quantitative assessment of the predators of *Nezara viridula* eggs and nymphs within a soybean agroecosystem using an ELISA. Environ. Entomol. 10: 402–405.

Ramseur, E. L., J. E. Box, and J. W. Johnson. 1989. Wheat growth and development, pp. 3–7. *In* Anonymous, Small grain resource handbook. Georgia Coop. Ext. Serv. Publ. COA-1: 1–70.

Rand, F. V. 1914. Some diseases of pecans. J. Agric. Res. 1: 303–338 + 5 plates.

Raney, H. G., and K. V. Yeargan. 1977. Seasonal abundance of common phytophagous and predaceous insects in Kentucky soybeans. Trans. Kentucky Acad. Sci. 38: 83–87.

Reich, R. C. (Ed.). 1991. Flue-cured tobacco field manual (3rd ed.). R. J. Reynolds Tob. Co., Winston-Salem, NC. 86 pp.

Reilly, C. C., and W. L. Tedders. 1989. Bacterial involvement in late season drop and kernel discolor of pecan. Proc. Annu. Conv. Southeastern Pecan Growers Assoc. 82: 195–199.

Rice, G. W. 1994. Pecans — A grower's perspective. Ag Press, Inc. Manhattan, Kansas. 198 pp.

Rice, R. E., J. K. Uyemoto, J. M. Ogawa, and W. M. Pemberton. 1985. New findings on pistachio problems. California Agric. 39: 15–18.

Rice, R. E., W. J. Bentley, and R. H. Beede. 1988. Insect and mite pests of pistachios in California. California Coop. Ext. Publ. 21452: 1–26.

Rider, D. A., and J. B. Chapin. 1992. Revision of the genus *Thyanta* Stål, 1862 (Heteroptera: Pentatomidae). II. North America, Central America, and the West Indies. J. New York Entomol. Soc. 100: 42–98.

Rider, D. A., and L. H. Rolston. 1995. Nomenclatural changes in the Pentatomidae (Hemiptera-Heteroptera). Proc. Entomol. Soc. Washington 97: 845–855.

Riley, C. V. 1882. Report of the entomologist, p. 138. U.S.D.A. Rep. Commissioner Agric. U. S. Government Printing Office, Washington, D.C. 214 pp.

Riley, C. V. 1885. Fourth Rep. U.S. Entomol. Comm., pp. 97–98. U. S. Government Printing Office, Washington, D.C. 399 pp.

Riley, D. G., G. K. Douce, and R. M. McPherson. 1997. Summary of losses from insect damage and costs of control in Georgia 1996. Georgia Agric. Exp. Stn. Spec. Publ. 91: 1–58.

Riley, T. J., J. Negron, and J. Baldwin. 1987. Southern green stink bug damage to field corn. Louisiana Agric. 30(3): 3,24.

Rinckner, C. M. 1991. Cultural and management practices. Alfalfa seed production and pest management. Western Reg. Ext. Publ. 12: 1–5.

Rings, R. W. 1955. Tarnished plant bug and stink bug injury to peaches. Proc. Ohio State Hortic. Soc. 108: 51–56.

Rings, R. W. 1957. Types and seasonal incidence of stink bug injury to peaches. J. Econ. Entomol. 50: 599–604.

Rings, R. W., and R. F. Brooks. 1958. Bionomics of the one-spot stink bug in Ohio. Ohio Agric. Exp. Stn. Res. Circ. 50: 1–16.

Roach, S. H. 1988. Stink bugs in cotton and estimation of damage caused by their feeding on fruiting structures. Proc. Beltwide Cotton Production Res. Conf., pp. 292–294.

Robinson, J. F., C. M. Smith, G. B. Trahan, and M. Hollay. 1980. Rice stink bug: relationship between adult infestation levels and damage. Louisiana Agric. Exp. Stn., Rice Exp. Stn., Annu. Prog. Rep. 72: 212–215.

Robinson, J. F., C. M. Smith, G. B. Trahan, and M. Hollay. 1981. Evaluation of 32 uniform rice nursery lines for rice stink bug resistance. Louisiana Agric. Exp. Stn., Rice Exp. Stn., Annu. Prog. Rep. 73: 278–285.

Robinson, J. F., C. M. Smith, G. B. Trahan, and R. M. Michot. 1987. Rice stink bug insecticide bioassay procedure. Louisiana Agric. Exp. Stn., Rice Exp. Stn., Annu. Prog. Rep. 79: 196–197.

Rolston, L. H. 1974. Revision of the genus *Euschistus* in Middle America (Hemiptera, Pentatomidae, Pentatomini). Entomol. Am. 48(1): 1–102.

Rolston, L. H., and R. L. Kendrick. 1961. Biology of the brown stink bug, *Euschistus servus* Say. J. Kansas Entomol. Soc. 34: 151–157.

Rolston, L. H., P. Rouse, and R. Mayes. 1966. Pecky rice. Arkansas Farm Res. 15(2): 6.

Rosenfeld, A. H. 1911. Insects and spiders in Spanish moss. J. Econ. Entomol. 4: 398–409.

Russell, E. E. 1952. Stink bugs on seed alfalfa in southern Arizona. U.S.D.A. Circ. 903: 1–19.

Russin, J. S., M. B. Layton, D. B. Orr, and D. J. Boethel. 1987. Within-plant distribution of, and partial compensation for, stink bug (Heteroptera: Pentatomidae) damage to soybean seeds. J. Econ. Entomol. 80: 215–220.

Russin, J. S., D. B. Orr, M. B. Layton, and D. J. Boethel. 1988a. Incidence of microorganisms in soybean seeds damaged by stink bug feeding. Phytopathology 78: 306–310.

Russin, J. S., M. B. Layton, D. B. Orr, and D. J. Boethel. 1988b. Stink bug damage to soybeans. Louisiana Agric. 32(1): 3, 24.

Ryan, M. A., and G. H. Walter. 1992. Sound communication in *Nezara viridula* (L.) (Heteroptera: Pentatomidae): further evidence that signal transmission is substrate-borne. Experientia 48: 1112–1115.

Ryker, T. C., and W. A. Douglas. 1938. Pecky rice investigations. Louisiana Agric. Exp. Stn. Biennial Rep., Rice Exp. Stn., 1937–1938, pp. 19–20.

Sailer, R. I. 1944. The genus *Solubea* (Heteroptera: Pentatomidae). Proc. Entomol. Soc. Washington 46: 105–127.

Sailer, R. I. 1952. A technique for rearing certain Hemiptera. U.S.D.A. Bur. Entomol. and Plant Quarantine ET-303: 1–5.

Sailer, R. I. 1953. A note on the bionomics of the green stink bug (Hemiptera: Pentatomidae). J. Kansas Entomol. Soc. 26: 70–71.

Sailer, R. I. 1954. Interspecific hybridization among insects with a report on crossbreeding experiments with stink bugs. J. Econ. Entomol. 47: 377–383.

Sane, I., D. R. Alverson, and J. W. Chapin. 1999. Efficiency of conventional sampling methods for determining arthropod densities in close-row soybeans. J. Agric. Urban Entomol. 16: 65–84.

Schalk, J. M., and R. L. Fery. 1982. Southern green stink bug and leaffooted bug: effect on cowpea production. J. Econ. Entomol. 75: 72–75.

Schalk, J. M., and R. L. Fery. 1986. Resistance in cowpea to the southern green stink bug. HortScience 21: 1189–1190.

Scheel, C. A., S. D. Beck, and J. T. Medler. 1957. Nutrition of plant-sucking Hemiptera. Science 125: 444–445.

Schoene, W. J., and G. W. Underhill. 1933. Economic status of the green stinkbug with reference to the succession of its wild hosts. J. Agric. Res. 46: 863–866.

Schotzko, D. J., and L. E. O'Keeffe. 1990a. Effect of pea and lentil development on reproduction and longevity of *Thyanta pallidovirens* (Stål) (Hemiptera: Heteroptera: Pentatomidae). J. Econ. Entomol. 83: 1333–1337.

Schotzko, D. J., and L. E. O'Keeffe. 1990b. Ovipositional rhythms of *Thyanta pallidovirens* (Hemiptera: Pentatomidae). Environ. Entomol. 19: 630–634.

Schueneman, T. J. 1993. An overview of Florida rice. Florida Coop. Ext. Serv. Fact Sheet AGR-64. http://hammock.ifas.ufl.edu/txt/fairs. 33 pp.

Schuh, R. T., and J. A. Slater. 1995. True bugs of the world (Hemiptera: Heteroptera). Classification and natural history. Cornell Univ. Press, Ithaca, NY. 336 pp.

Schumann, F. W., and J. W. Todd. 1982. Population dynamics of the southern green stink bug (Heteroptera: Pentatomidae) in relation to soybean phenology. J. Econ. Entomol. 75: 748–753.

Schuster, D. J. 1991. Insect control on fresh market tomatoes in west-central Florida, fall 1988. Insecticide and Acaricide Tests 16: 115.

Schuster, D. J. 1998. Insect management on fresh market tomatoes, spring 1997A. Arthropod Manage. Tests 23: 158–159.

Schuster, D. J., and J. E. Polston. 1998a. Management of whiteflies, bugs, and leafminers on fresh market tomatoes, fall 1996. Arthropod Manage. Tests 23: 156–157.

Schuster, D. J., and J. E. Polston. 1998b. Insect management on fresh market tomatoes, spring 1997B. Arthropod Manage. Tests 23: 159–160.

Scott, C. E., and H. F. Madsen. 1950. Stink-bug injury to pears. Western Fruit Grower 4(12): 16.

Scudder, G. G. E., and D. B. Thomas, Jr. 1987. The green stink bug genus Chlorochroa Stål (Hemiptera: Pentatomidae) in Canada. Can. Entomol. 119: 83–93.

Sedlacek, J. D., and L. H. Townsend. 1988. Impact of *Euschistus servus* and *E. variolarius* (Heteroptera: Pentatomidae) feeding on early growth stages of corn. J. Econ. Entomol. 81: 840–844.

Seymour, J. E., and G. J. Bowman. 1994. Russet coloration in *Nezara viridula* (Hemiptera: Pentatomidae): an unreliable indicator of diapause. Environ. Entomol. 23: 860–863.

Shearer, P. W., and V. P. Jones. 1996a. Diel feeding pattern of adult female southern green stink bug (Hemiptera: Pentatomidae). Environ. Entomol. 25: 599–602.

Shearer, P. W., and V. P. Jones. 1996b. Suitability of macadamia nut as a host plant of *Nezara viridula* (Hemiptera: Pentatomidae). J. Econ. Entomol. 89: 996–1003.

Shepard, B. M., K. D. Elsey, A. E. Muckenfuss, and H. D. Justo, Jr. 1994. Parasitism and predation on egg masses of the southern green stink bug, *Nezara viridula* (L.) (Heteroptera: Pentatomidae), in tomato, okra, cowpea, soybean, and wild radish. J. Agric. Entomol. 11: 375–381.

Shepard, M., G. R. Carner, and S. G. Turnipseed. 1977. Colonization and resurgence of insect pests of soybean in response to insecticides and field isolation. Environ. Entomol. 6: 501–506.

Simmons, A. M., and K. V. Yeargan. 1988a. Feeding frequency and feeding duration of the green stink bug (Hemiptera: Pentatomidae) on soybean. J. Econ. Entomol. 81: 812 815.

Simmons, A. M., and K. V. Yeargan. 1988b. Development and survivorship of the green stink bug, *Acrosternum hilare* (Hemiptera: Pentatomidae) on soybean. Environ. Entomol. 17: 527–532.

Simmons, A. M., and K. V. Yeargan. 1990. Effect of combined injuries from defoliation and green stink bug (Hemiptera: Pentatomidae) and influence of field cages on soybean yield and seed quality. J. Econ. Entomol. 83: 599–609.

Slater, J. A., and R. M. Baranowski. 1978. How to know the true bugs (Hemiptera—Heteroptera). Wm. C. Brown Co. Publ., Dubuque, IA. 256 pp.

Smilanick, J. M., F. G. Zalom, and L. E. Ehler. 1996. Effect of methamidophos residue on the pentatomid egg parasitoids *Trissolcus basalis* and *T. utahensis* (Hymenoptera: Scelionidae). Biol. Control 6: 193–201.

Smith, C. M., J. L. Bagent, S. D. Linscombe, and J. F. Robinson. 1986. Insect pests of rice in Louisiana. Louisiana Agric. Exp. Stn. Bull. 774: 1–24.

Smith, K. 1994. Importance of soybeans, p. 3. *In* L. G. Higley and D. J. Boethel (Eds.), Handbook of soybean insect pests. Entomol. Soc. Am. Publ., Lanham, MD. 136 pp.

Smith, M. T. 1996a. Trap cropping system for control of hemipteran pests in pecan: a progress report. Proc. Southeastern Pecan Growers Assoc. 89: 71–81.

Smith, M. T. 1996b. Trap cropping system: for control of stink bugs. The Pecan Grower 8(2): 6–9.

Smith, M. T. 1999. Managing kernel feeding hemipterans in pecan, pp. 163–166. *In* B. McCraw, E. H. Dean, and B. W. Wood (Eds.), Pecan industry: current situation and future challenges, third national pecan workshop proceedings. U.S.D.A. Agric. Res. Serv. 1998–04. 234 pp.

Snapp, O. I. 1941. Insect pests of the peach in the eastern states. U.S.D.A. Farmers' Bull. 1861: 1–34.

Snapp, O. I. 1954. Insect pests of the peach in the eastern states. U.S.D.A. Farmers' Bull. 1861: 1–32.

Sorenson, C. J., and E. W. Anthon. 1936. Preliminary studies of *Acrosternum hilaris* (Say) in Utah orchards. Utah Acad. Sci., Arts, and Letters 13: 229–232.

Southwick, L. M., J. Yanes, Jr., D. J. Boethel, and G. H. Willis. 1986. Leaf residue compartmentalization and efficacy of permethrin applied to soybean. J. Entomol. Sci. 21: 248–253.

Sparks, A. N. and E. R. Mitchell. 1979. Economic thresholds of *Heliothis* species on corn, pp. 51–56. *In* W. L. Sterling (Chair), Economic thresholds and sampling of *Heliothis* species on cotton, corn, soybeans and other host plants. Southern Coop. Ser. Bull. 231: 1–159.

Spink, W. T. 1960. Summary of insect conditions in Louisiana for 1960, pp. 51–58. *In* W. T. Spink (Ed.), Insect conditions in Louisiana 1960. Louisiana State University 3: 1–58.

Staddon, B. W. 1979. The scent glands of Heteroptera, pp. 351–418. *In* J. E. Treherne, M. J. Berridge, and V. B. Wigglesworth (Eds.), Advances in insect physiology. Vol. 14. Acad. Press, New York, NY. 440 pp.

Stam, P. A., L. D. Newsom, and E. N. Lambremont. 1987. Predation and food as factors affecting survival of *Nezara viridula* (L.) (Hemiptera: Pentatomidae) in a soybean eco-system. Environ. Entomol. 16: 1211–1216.

Stein, L. A. 1985. Stink bugs: proper management can control kernel spots, black pit. Pecan South 19(5): 15–18, 37.

Stone, A., C. W. Sabrosky, W. W. Wirth, R. H. Foote, and J. R. Coulson. 1965. A catalog of the Diptera of America north of Mexico. U.S.D.A. Agric. Res. Serv., Agric. Handbook 276. U. S. Government Printing Office, Washington, D. C. 1,696 pp.

Stoner, D. 1922. Report on the Scutelleroidea collected by the Barbados-Antiqua expedition from the University of Iowa in 1918. Univ. Iowa Stud. Nat. Hist. 10: 3–17 (+ 3 plates).

Streams, F. A., and D. Pimentel. 1963. Biology of the harlequin bug, *Murgantia histrionica.* J. Econ. Entomol. 56: 108–109.

Strickland, E. H. 1953. An annotated list of the Hemiptera (S.L.) of Alberta. Can. Entomol. 85: 193–214.

Stultz, H. T. 1955. The influence of spray programs on the fauna of apple orchards in Nova Scotia. VIII. Natural enemies of the eye-spotted bud moth, *Spilonota ocellana* (D. & S.) (Lepidoptera: Olethreutidae). Can. Entomol. 87: 79–85.

Sudarsono, H., J. L. Bernhardt, and N. P. Tugwell. 1992. Survival of immature *Telenomus podisi* (Hymenoptera: Scelionidae) and rice stink bug (Hemiptera: Pentatomidae) embryos after field applications of methyl parathion and carbaryl. J. Econ. Entomol. 85: 375–378.

Sullivan, M. J., and C. H. Brett. 1974. Resistance of commercial crucifers to the harlequin bug in the Coastal Plain of North Carolina. J. Econ. Entomol. 67: 262–264.

Summerfield, R. J., J. S. Pate, E. H. Roberts, and H. C. Wien. 1985. The physiology of cowpeas, pp. 65–101. *In* S. R. Singh and K. O. Rachie (Eds.), Cowpea research, production and utilization. John Wiley & Sons, New York, NY. 460 pp.

Swan, L. A., and C. S. Papp. 1972. The common insects of North America. Harper & Row, Publ., New York, NY. 750 pp.

Swanson, M. 1960. Effect of rice stink bug infestations on yield and quality of magnolia rice, pp. 12–13. *In* W. T. Spink (Ed.), Insect conditions in Louisiana 1960. Louisiana State University 3: 1–58.

Swanson, M. C., and L. D. Newsom. 1962. Effect of infestation by the rice stink bug, *Oebalus pugnax*, on yield and quality in rice. J. Econ. Entomol. 55: 877–879.

Taylor, J. D., C. S. Bundy, M. L. Wells, and R. M. McPherson. 1998. Control of stink bugs and lepidopterans on Georgia soybeans, 1997. Arthropod Manage. Tests 23: 284–285.

Teare, I. D., and H. F. Hodges. 1994. Soybean ecology and physiology, pp. 4–7. *In* L. G. Higley and D. J. Boethel (Eds.), Handbook of soybean insect pests. Entomol. Soc. Am. Publ., Lanham, MD. 136 pp.

Telford, A. D. 1957. Arizona cotton crops. Arizona Agric. Exp. Stn. Bull. 286: 1–60.

Temerak, S. A., and W. H. Whitcomb. 1984. Parasitoids of predaceous and phytophagous pentatomid bugs in soybean fields at two sites of Alachua County, Florida. Zeitschr. Angew. Entomol. 97: 279–282.

Texas Agricultural Extension Service. 1997. Insect management alternatives — rice stink bug, pp. 42–47. *In* 1997 rice production guidelines. Texas Agric. Ext. Serv. Publ. D-1253: 1–65.

Thomas, D. B., Jr. 1983. Taxonomic status of the genera *Chlorochroa* Stål, *Rhytidilomia* [sic] Stål, *Liodermion* Kirkaldy, and *Pitedia* Reuter, and their included species (Hemiptera: Pentatomidae). Ann. Entomol. Soc. Am. 76: 215–224.

Thomas, G. D., C. M. Ignoffo, C. E. Morgan, and W. A. Dickerson. 1974. Southern green stink bug: influence on yield and quality of soybeans. J. Econ. Entomol. 67: 501–503.

Todd, J. W. 1976. Effects of stink bug feeding on soybean seed quality. World Soybean Res., pp. 611–618.

Todd, J. W. 1981. Effects of stinkbug damage on soybean quality. Proc. Internatl. Congr. on Soybean Seed Quality and Stand Establishment 22: 46–51.

Todd, J. W. 1989. Ecology and behavior of *Nezara viridula*. Annu. Rev. Entomol. 34: 273–292.

Todd, J. W., and D. C. Herzog. 1980. Sampling phytophagous Pentatomidae on soybean, pp. 438–478. *In* M. Kogan and D. C. Herzog (Eds.), Sampling methods in soybean entomology. Springer-Verlag, New York, NY. 587 pp.

Todd, J. W., and W. J. Lewis. 1976. Incidence and oviposition patterns of *Trichopoda pennipes* (F.), a parasite of the southern green stink bug, *Nezara viridula* (L.). J. Georgia Entomol. *Soc.* 11: 50–54.

Todd, J. W., and B. G. Mullinix. 1985. Effects of insect-pest complexes on soybean, pp. 624–634. World Soybean Research Conference III: Proceedings. R. Shibles (Ed.). Westview Press, Boulder, CO. 1262 pp.

Todd, J. W., and F. W. Schumann. 1988. Combination of insecticide applications with trap crops of early maturing soybean and southern peas for population management of *Nezara viridula* in soybean (Hemiptera: Pentatomidae). J. Entomol. Sci. 23: 192–199.

Todd, J. W., and S. G. Turnipseed. 1974. Effects of southern green stink bug damage on yield and quality of soybeans. J. Econ. Entomol. 67: 421–426.

Todd, J. W., and H. Womack. 1973. Secondary infestations of cigarette beetle in soybean seed damaged by southern green stink bug. Environ. Entomol. 2: 720.

Todd, J. W., M. D. Jellum, and D. B. Leuck. 1973. Effects of southern green stink bug damage on fatty acid composition of soybean oil. Environ. Entomol. 2: 685–689.

Todd, J. W., R. M. McPherson, and D. J. Boethel. 1994. Management tactics for soybean insects, pp. 115–117. *In* L. G. Higley and D. J. Boethel (Eds.), Handbook of soybean insect pests. Entomol. Soc. Am. Publ., Lanham, MD. 136 pp.

Tonhasca, A., Jr., and B. R. Stinner. 1991. Effects of strip intercropping and no-tillage on some pests and beneficial invertebrates of corn in Ohio. Environ. Entomol. 20: 1251–1258.

Torre-Bueno, J. R., de la. 1912. *Nezara viridula* Linné, an hemipteron new to the northeastern United States. Entomol. News 23: 316–318.

Toscano, N. C., and V. M. Stern. 1976a. Cotton yield and quality loss caused by various levels of stink bug infestations. J. Econ. Entomol. 69: 53–56.

Toscano, N. C., and V. M. Stern. 1976b. Dispersal of *Euschistus conspersus* from alfalfa grown for seed to adjacent crops. J. Econ. Entomol. 69: 96–98.

Toscano, N. C., and V. M. Stern. 1976c. Development and reproduction of *Euschistus conspersus* at different temperatures. Ann. Entomol. Soc. Am. 69: 839–840.

Toscano, N. C., and V. M. Stern. 1980. Seasonal reproductive condition of *Euschistus conspersus*. Ann. Entomol. Soc. Am. 73: 85–88.

Townsend, L. H., and J. D. Sedlacek. 1986. Damage to corn caused by *Euschistus servus*, *E. variolarius*, and *Acrosternum hilare* (Heteroptera: Pentatomidae) under greenhouse conditions. J. Econ. Entomol. 79: 1254–1258.

Tugwell, N. P. 1986. Pecky rice control. Texas Agric. Exp. Stn., Proc. 21st Rice Tech. Working Group, pp. 89–90.

Tugwell, P., E. P. Rouse, and R. G. Thompson. 1973. Insects in soybeans and a weed host (*Desmodium* sp.). Arkansas Agric. Exp. Stn. Rep. Ser. 214: 1–18.

Turner, W. F. 1918. *Nezara viridula* and kernel spot of pecan. Science (N.S.) 47: 490–491.

Turner, W. F. 1923. Kernel spot of pecan caused by the southern green soldier bug. J. Econ. Entomol. 16: 440–445.

Turnipseed, S. G., and M. Kogan. 1976. Soybean entomology. Annu. Rev. Entomol. 21: 247–282.

Turnipseed, S. G., and M. Kogan. 1983. Soybean pests and indigenous natural enemies, pp. 1–6. *In* H. N. Pitre (Ed.), Natural enemies of arthropod pests in soybean. South Carolina Agric. Exp. Stn. (Southern Coop. Ser.) Bull. 285: 1–90.

Turnipseed, S. G., and M. J. Sullivan. 1976. Plant resistance in soybean insect management, pp. 549–560. *In* L. D. Hill (Ed.), World soybean research. Interstate Printers and Publ., Inc., Danville, IL. 1,073 pp.

Turnipseed, S. G., M. J. Sullivan, J. E. Mann, and M. E. Roof. 1995. Secondary pests in transgenic Bt cotton in South Carolina. Proc. 1995 Beltwide Cotton Conf. 2: 768–769.

Twinn, C. R. 1938. A summary of the insect pest situation in Canada in 1938. Annu. Rep. Entomol. Soc. Ontario 69: 121–134.

Twinn, C. R. 1939. A summary statement in regard to some of the more important insect pests in Canada in 1939. Annu. Rep. Entomol. Soc. Ontario 70: 115–125.

Underhill, G. W. 1934. The green stinkbug. Virginia Agric. Exp. Stn. Bull. 294: 1–26.

U.S.D.A. National Agricultural Statistics Service. 1998. U.S.D.A. Agric. Stat. 1998, Washington, D. C. 514 pp.

Vergara, B. S. 1991. Rice plant growth and development, pp. 13–22. *In* Bor S. Luh (Ed.), Rice production (Vol. I). Van Nostrand Reinhold, New York, NY. 439 pp.

Velasco, L. R. I., and G. H. Walter. 1992. Availability of different host plant species and changing abundance of the polyphagous bug *Nezara viridula* (Hemiptera: Pentatomidae). Environ. Entomol. 21: 751–759.

Velasco, L. R. I., and G. H. Walter. 1993. Potential of host-switching in *Nezara viridula* (Hemiptera: Pentatomidae) to enhance survival and reproduction. Environ. Entomol. 22: 326–333.

Velez, J. R. 1974. Observaciones sobre la biologia de la chinche verde, *Nezara viridula* (L.), en el valle del fuerte sin. Folia Entomol. Mexico 28: 5–12.

Viator, H. P., and C. M. Smith. 1980. Effects of stinkbug feeding activity on soft red winter wheat quality. Louisiana Agric. Exp. Stn. Rep. Projects for 1980, Dept. Agron., pp. 38–39.

Viator, H. P., A. Pantoja-Lopez, and C. M. Smith. 1982. Effects of rice [*Oebalus pugnax* (F.)] and southern green [*Nezara viridula* (L.)] stink bug damage on wheat seed yield and quality. Louisiana Agric. Exp. Stn. Rep. Projects for 1982, Dept. Agron., pp. 40–141.

Viator, H. P., A. Pantoja, and C. M. Smith. 1983. Damage to wheat seed quality and yield by the rice stink bug and southern green stink bug (Hemiptera: Pentatomidae). J. Econ. Entomol. 76: 1410–1413.

Vigna Crop Germplasm Committee. 1996. *Vigna* germplasm. Current status and future needs. http://www.ars-grin.gov/npgs/cgc_reports/vigna96.html. 7 pp.

Walgenbach, J. F., and C. R. Palmer. 1998. Insect control on staked tomatoes, 1997. Arthropod Manage. Tests 23: 166–167.

Walgenbach, J. F., and C. R. Palmer. 1999a. Insect control on staked tomatoes, 1998. Arthropod Manage. Tests 24: 184–185.

Walgenbach, J. F., and C. R. Palmer. 1999b. Insect control on staked tomatoes using imidacloprid (Admire), 1998. Arthropod Manage. Tests 24: 185.

Walker, H. G., and L. D. Anderson. 1933. Report on the control of the harlequin bug, *Murgantia histrionica* Hahn, with notes on the severity of an outbreak of this insect in 1932. J. Econ. Entomol. 26: 129–135.

Walsh, B. D. 1866. The Texan cabbage-bug. (*Strachia histrionica* Hahn.). Practical Entomol. 1: 110.

Wang, Q., and J. G. Millar. 1997. Reproductive behavior of *Thyanta pallidovirens* (Heteroptera: Pentatomidae). Ann. Entomol. Soc. Am. 90: 380–388.

Waterhouse, D. F., D. A. Forss, and R. H. Hackman. 1961. Characteristic odour components of the scent of stink bugs. J. Insect Physiol. 6: 113–121.

Watson, J. R. 1918. Insects of a citrus grove. Florida Agric. Exp. Stn. Bull. 148: 165–267.

Watson, J. R. 1919. Florida truck and garden insects. Florida Agric. Exp. Stn. Bull. 151: 111–211.

Watson, J. R. 1934. Notes on *Nezara viridula* (L.). Florida Entomol. 18: 43.

Watson, J. R., and A. N. Tissot. 1942. Insects and other pests of Florida vegetables. Florida Agric. Exp. Stn. Bull. 370: 1–118.

Way, M. O. 1984. Insect management for Lemont, p. 35. *In* N. P. Clarke (Dir.), The semidwarfs — a new era in rice production. Texas Agric. Exp. Stn. B-1462: 1–44.

Way, M. O. 1990. Insect pest management in rice in the United States, pp. 181–189. *In* B. T. Grayson, M. B. Green, and L. G. Copping (Eds.), Pest management in rice. Elsevier Applied Science, New York, NY. 536 pp.

Way, M. O. 1994. Status of soybean insect pests in the United States, pp. 15–16. *In* L. G. Higley and D. J. Boethel (Eds.), Handbook of soybean insect pests. Entomol. Soc. Am. Publ., Lanham, MD. 136 pp.

Way, M. O., and R. G. Wallace. 1986a. 'Peck' in relation to the seasonal distribution of the rice stink bug on Labelle and Lemont. Texas Agric. Exp. Stn., Proc. 21st Rice Tech. Working Group, p. 89.

Way, M. O., and R. G. Wallace. 1986b. Field applied insecticides for rice stink bug control. Texas Agric. Exp. Stn. Publ. PR-4351: 1–4.

Way, M. O., and R. G. Wallace. 1990. Residual activity of selected insecticides for control of rice stink bug (Hemiptera: Pentatomidae). J. Econ. Entomol. 83: 591–595.

Way, M. O. and R. G. Wallace. 1995. Evaluation of Karate for management of arthropods in soybean, 1992. Arthropod Management Tests 20: 246–247.

Way, M. O., C. C. Bowling, and R. G. Wallace. 1987. Initial and residual activity of insecticides for control of rice stink bug, *Oebalus pugnax* (Fabricius). Texas Agric. Exp. Stn. PR-4504: 1–8.

Webb, J. L. 1920. How insects affect the rice crop. U.S.D.A. Farmers' Bull. 1086: 1–11.

Weed Science Society of America. 1984. Composite list of weeds. J. Weed Sci. Soc. Am. 32 (Supplement 2): 1–137.

Wene, G. P., and L. W. Sheets. 1964. Notes on and control of stink bugs affecting cotton in Arizona. J. Econ. Entomol. 57: 60–62.

White, W. H., and L. W. Brannon. 1933. The harlequin bug and its control. U.S.D.A. Farmers' Bull. 1712: 1–10.

Whitmarsh, R. D. 1914. The green soldier bug (*Nezara hilaris*). J. Econ. Entomol. 7: 336–339.

Whitmarsh, R. D. 1917. The green soldier bug. *Nezara hilaris* Say. Order, Hemiptera. Family, Pentatomidae. A recent enemy in northern Ohio peach orchards. Ohio Agric. Exp. Stn. Bull. 310: 519–552.

Wier, A. T., D. J. Boethel, and J. S. Mink. 1991. Control of southern green stink bug on soybean, 1990. Insecticide and Acaricide Tests 16: 225–226.

Wilde, G. E. 1968. A laboratory method for continuously rearing the green stink bug. J. Econ. Entomol. 61: 1763–1764.

Wilde, G. E. 1969. Photoperiodism in relation to development and reproduction in the green stink bug. J. Econ. Entomol. 62: 629–630.

Wilks, J. M. 1964. The spined stink bug: cause of cottony spot in pear in British Columbia. Can. Entomol. 96: 1198–1201.

Williams, M. R. 1995. Beltwide cotton insect losses 1994. Proc. 1995 Beltwide Cotton Conf. 2: 746–757.

Wiseman, B. R., and W. W. McMillian. 1971. Damage to sorghum in south Georgia by Hemiptera. J. Georgia Entomol. Soc. 6: 237–242.

Wood, B. W., and W. L. Tedders. 1996. Stink bug feeding costs massive nut dropping in orchards. The Pecan Grower 8(2): 14.

Woodside, A. M. 1946a. Life history studies of *Euschistus servus* and *E. tristigmus*. J. Econ. Entomol. 39: 161–163.

Woodside, A. M. 1946b. Cat-facing and dimpling in peaches. J. Econ. Entomol. 39: 158–161.

Woodside, A. M. 1947. Weed hosts of bugs which cause cat-facing of peaches in Virginia. J. Econ. Entomol. 40: 231–233.

Woodside, A. M. 1949. Tests of insecticides for control of cat-facing on peaches. J. Econ. Entomol. 42: 335–338.

Woodside, A. M. 1950. Cat-facing and dimpling of peaches. Virginia Agric. Exp. Stn. Bull. 435: 1–18.

Wyffels Hybrids Inc. 1998. Late-season corn development. http://www.wyffels.com/corndev3.htm. 5 pp.

Yates, I. E., W. L. Tedders, and D. Sparks. 1991. Diagnostic evidence of damage on pecan shells by stink bugs and coreid bugs. J. Am. Soc. Hortic. Sci. 116: 42–46.

Yeargan, K. V. 1977. Effects of green stink bug damage on yield and quality of soybeans. J. Econ. Entomol. 70: 619–622.

Yeargan, K. V. 1979. Parasitism and predation of stink bug eggs in soybean and alfalfa fields. Environ. Entomol. 8: 715–719.

Yeargan, K. V. 1985. Alfalfa: status and current limits to biological control in the eastern U. S., pp. 521–536. *In* M. Hoy and D. Herzog (Eds.), Biological control in agricultural IPM systems. Acad. Press, Inc., New York, NY. 589 pp.

Yeargan, K. V., J. W. Todd, A. J. Mueller, and W. C. Yearian. 1983. Dynamics of natural enemy populations, pp. 32–38. *In* H. N. Pitre (Ed.), Natural enemies of arthropod pests in soybean. South Carolina Agric. Exp. Stn. (Southern Coop. Ser.) Bull. 285: 1–90.

Yonce, C., and R. Mizell, III. 1997. Stink bug trapping with a pheromone. Proc. Southeastern Pecan Growers Assoc. 90: 54–56.

Yonke, T. R. 1991. Order Hemiptera, pp. 22–65. *In* F. W. Stehr (Ed.), Immature insects, Volume 2. Kendall/Hunt Publ. Co., Dubuque, IA. 975 pp.

Young, W. R., and G. L. Teetes. 1977. Sorghum entomology. Ann. Rev. Entomol. 22: 193–218.

Youther, M. L., and J. E. McPherson. 1975. A study of fecundity, fertility, and hatch in *Euschistus servus* (Hemiptera: Pentatomidae) with notes on precopulatory and copulatory behavior. Trans. Illinois State Acad. Sci. 68: 321–338.

Zalom, F. G., and J. S. Zalom. 1992. Stink bugs in California tomatoes. California Tomato Grower 35(7): 8,10, 11.

Zalom, F. G., J. M. Smilanick, and L. E. Ehler. 1997a. Fruit damage by stink bugs (Hemiptera: Pentatomidae) in bush-type tomatoes. J. Econ. Entomol. 90: 1300–1306.

Zalom, F. G., J. M. Smilanick, and L. E. Ehler. 1997b. Spatial pattern and sampling of stink
    bugs (Hemiptera: Pentatomidae) in processing tomatoes, pp. 75–79. *In* G. A. Maciel,
    G. M. B. Lopes, C. Hayward, R. R. L. Mariano, and E. A. de A. Maranhao (Eds.),
    Proceedings of the 1st International Conference on the Processing Tomato. ASHS Press,
    Alexandria, VA. 190 pp.
Zeiss, M. R., and T. H. Klubertanz. 1994. Sampling programs for soybean arthropods,
    pp. 539–601. *In* L. P. Pedigo and G. D. Buntin (Eds.), Handbook of sampling methods
    for arthropods in agriculture. CRC Press, Inc., Boca Raton, FL. 714 pp.

Index